S0-AEB-059

Think About the Future

We are all busy people. Let's face it, we don't have the time or the desire to climb smokestacks or confront whaling vessels. But there are lots of things we can do differently every day. Without effort and with very little thought, we can all make a difference to our planet Earth.

The way we use paper, how we use water, clean ourselves and our homes, what we do in our gardens, our cars, at work and at school, all this determines what kind of planet we inhabit. We owe it to our children to give them the greenest, healthiest Earth possible as well as the tools, the knowledge and the wisdom to carry on the process in the future.

In only two minutes a day, we can change the direction in which we are all headed . . .

TWO MINUTES A DAY FOR A GREENER PLANET

QUICK AND SIMPLE THINGS YOU CAN DO TO SAVE OUR EARTH

MARJORIE LAMB

HarperPaperbacks
A Division of HarperCollins*Publishers*

HarperPaperbacks *A Division of* HarperCollins*Publishers*
10 East 53rd Street, New York, N.Y. 10022

Copyright © 1990 by Marjorie Lamb
All rights reserved. No part of this book may be used or
reproduced in any manner whatsoever without written
permission of the publisher, except in the case of brief
quotations embodied in critical articles and reviews. For
information address HarperCollins*Publishers*,
10 East 53rd Street, New York, N.Y. 10022.

A trade paperback edition of this book was published in
1990 by Harper & Row, Publishers, Inc.

HarperCollins*Publishers* will facilitate planting two trees
for every one tree used in the manufacturing of this book
with an affiliate of the Rainforest Action Network.

Cover photo courtesy of NASA

First HarperPaperbacks printing: April 1991

Printed in the United States of America

HarperPaperbacks and colophon are trademarks of
HarperCollins*Publishers*

10 9 8 7 6 5 4 3 2 1

Publisher's Note

For every tree used to print this book, we are planting two.

Traditionally, publishers who cared about the environment have printed on recycled paper. Unfortunately, problems created by the bleaches and chemicals used in the recycling process, and by the lack of book-quality recycled paper, make this an imperfect answer.

Restoring balance to the environment by planting two trees to replace every one we use is a positive alternative. And on the advice of experts, the cover stock of this book is protected with a "press varnish," which is biodegradable, unlike the usual plastic film coating.

Even before Marjorie's book reached our office, we had established a recycling program at Harper & Row San Francisco, and implemented a tree-planting alternative on other books we publish. We've already turned a bit green, but that doesn't mean we didn't have much to learn from this book. The beauty of *2 Minutes a Day for a Greener Planet*, and the reason we're so pleased to be publishing it, is that it speaks to everybody. Whether you're concerned but don't know what you can do, or are already active and looking for ways to be even more green, Marjorie provides ideas for action. Each person's actions make a difference. It is the good spiritedness and creativity of this book that make it so valuable.

We are indebted to our colleagues at HarperCollins Canada, for initiating the publication of this book.

To Caroline and Barry,
enthusiastic recyclers, willing walkers,
scrupulous savers, skeptical consumers,
ardent composters, and loving supporters
of all my endeavors.

Acknowledgments

Thanks to everyone who supported and encouraged me, shared their valuable information and insights, and contributed to the completion of the project, particularly:

The superb producers and crew of CBC Radio's *Metro Morning*, who had the instinct to know that people want to help the environment, and helped me develop the "Two Minute Ecologist" into a regular report: Ken Wolf, Marieke Meyer, Nancy Boyle and Joe Coté.

The president of HarperCollins, Stanley Colbert, who heard me on CBC Radio, phoned and asked, "Do you want to write a book?" Without hesitation I said, "Yes," and started to write that day.

My lawyer, Heather Mitchell, who tackles the world for me and gives me excellent advice on everything from the environment to my wardrobe.

The editor-in-chief of HarperCollins, David Colbert, who was ruthless in his advice and opinions, generous with his encouragement, and who put in long hours of overtime to get the book done sooner and better.

My editor, Marq de Villiers, who kept me in line with

his incisive comments and taught me to write without exclamation points!

The many hard-working environmentalists who answered my calls promptly and gave me much needed information: Trudy Richards, Greenpeace; Gudrun Knoessl and Jill Dunkley, Recycling Council of Ontario; Julia Langer, Friends of the Earth; Mary Perlmutter, Canadian Organic Growers; Gerard Coffey, Pollution Probe; Don Huff, Forests for Tomorrow; Ian Kirkham, Federation of Ontario Naturalists; Monte Hummel, Pegi Dover and Alison Wering, World Wildlife Fund; Linda Varangu, Ontario Waste Exchange; Paul Griss, Canadian Nature Federation; and Jim Richardson, Sierra Club.

The producers and crew of CBC Television's *Midday,* who gave me the opportunity to broadcast environmental advice nationally: Sophia Hadzepetros, Valerie Pringle and Ralph Benmurgi.

The staff of Harper Collins, who worked day and night, weekends included: Danielle Koch, Kathryn Schroeder, Kevin Hanson, Laura Krakowec, Paulette Burt, Donna Dunlop and freelancer Melanie Colbert.

Nancy Colbert, who supported and encouraged me, fed me information, and spread her cheerfulness generously.

V. Tony Hauser, for my photograph.

My family. You're all in the book. There wouldn't be a book without you.

Table of Contents

Introduction

Some days I think that Henny Penny is the news editor of every newspaper and television network in the country. You can't pick up a paper or a magazine or turn on the news without hearing that the sky is falling in, that our planet is going to hell in a handbasket, that our environment is so polluted, contaminated, depleted, overworked and out of order that we've only got five or 10 good years left to fix everything.

Some of this talk is practical and sensible. But a lot of it sounds like it's coming from somebody else's planet altogether. We hear phrases like "ozone depletion," "greenhouse effect," "toxic dumps" and "nonrenewable resources." We know they are the dirty words of today. But none of them hit home, and none of them seem like anything we could fix. What can we do about a hole in the ozone over the Antarctic, global warming and the destruction of rainforests? We've got three loads of laundry to finish before going back to the office.

**Most of the tips in this book will cost you
less than two minutes to put into practice.
But they will last you a lifetime.**

I've always believed in the great benefits offered by
people who make small, personal commitments to the
environment. My articles, radio reports and television ap-
pearances have focused on what a person can do to help
the planet by giving just a few minutes each day. When
I started, I hoped people were listening. But I wasn't
prepared for the avalanche of enthusiastic responses from
people who were working toward the same goal in the
same easy way. So many letters say, "Here's what we
do" or "Here's a quick and easy idea to help the environment." People everywhere want to help. They understand that the way we use paper, how we use water,
how we clean ourselves and our homes, what we do in
our gardens, our cars, at work and at school all determine
what kind of a planet we inhabit.

**The state of the environment has every-
thing to do with you and me.**

And by helping to restore health to our planet, we can
reconnect ourselves and our lives with the Earth in ways
that can bring us great joy.

Even though I grew up as part of the Baby Boom
generation, I learned my habits from my parents, who
came from the Depression generation. Some of the lessons from the past: Forego buying new clothes, and patch
the old. Grow your own food crops, and put up preserves.
Find uses for every little scrap of wood or fabric or paper
or metal. Buy less and repair more. "Make it do, wear it
out, use it up."

The necessity of thriftiness, impressed upon me in childhood, has become a virtue in the Age of the Environment. So even though we want to help the environment, the bonus is we could save quite a bit of money, just by following some of the advice in this book.

You'll be surprised to find that doing good things for the environment is habit forming. And one day, you might even find yourself deciding to ride your bike to work for a change, hang your clothes on the clothesline instead of putting them in the dryer, or forego the excessively packaged TV dinner, because it suddenly strikes you that each of those decisions is a little step towards a greener planet.

I have a daughter. Before I became a parent, I never thought I could ever love anyone so intensely. Not long after she was born, we started hearing more and more of those "dirty words" on the news, and I started to get scared. What kind of a world would my daughter inherit? I started to feel guilty that I was so busy with my home and office life I was neglecting her larger needs for a clean, safe planet.

But I didn't relish the thoughts of whisking her off to some secluded part of the world where she'd be safe from any and all possible harm—even supposing that such a spot could be found. I wanted her to have a "normal" life. I decided we couldn't wait for politicians and governments and businesses to act. We had to start doing something right here, right now, at home.

You'll be reading about my daughter, Caroline, from time to time in this book, and also about my partner, Barry, and my parents and other members of my family. Everything I write about here is stuff that's relevant to our everyday life. I try to follow my own advice, since so much of it comes from experience.

This is not advice to corporations or governments. It's not advice to professors or environmentalists.

> **It's just some tips and common sense that ordinary, everyday people can put into practice as they go about their daily lives.**

I don't have to tell you what shape the world is in. The doom-sayers are doing a pretty good job of that. But I'm willing to bet that you'll find more than a few ideas in this book that will give you the incentive to start acting positively to change the direction in which we're headed. You'll start to feel good about yourself when you hear people talk about the environment.

We're all busy people. Most of us don't have the time or even the desire to climb smokestacks or confront whaling vessels. Sometimes we can't even find the time to write letters to our politicians. But there are lots of things that you and I do every day that we could do differently, without much effort, and with just a little thought. Nearly all the things I discuss here are so easy to put into practice that making habits out of them will be simple.

I want my daughter to have a special life. I want her to have clean air to breathe, clean water to drink, healthy food to eat and a green planet to enjoy. We all want the same for our children, our grandchildren, and all people of the world. What could be more important? If all of us do a little bit every day, together we can make a world of difference to our planet Earth.

M.L.
January, 1990
Toronto

1

SAVING
WATER

Why Save Something That Covers
Two-thirds Of The Earth?

All life on this planet is supported by a fixed quantity of
water. We use the same water over and over again, the
same water which our grandparents used for brick-
making, the same water in which Shakespeare washed
his feet, the same water in which Moses floated in a basket
through the bullrushes, the same water the ancient Romans transported through their aqueducts to support life
in their city. In fact, the water that you used to brush your
teeth this morning is over four billion years old. So have
a little respect.

> **Of all water on our planet Earth, 97% is
> salt water. Only 3% is fresh water, and
> most of that is frozen in the polar ice caps.**

Less than 1% of Earth's water is available for our use.

We can't make new water, any more than we can make new land. If we misuse the water we have, we can't send out for some fresh stuff. Water comes out of the tap in unlimited quantities whenever we want it. We generally assume that we have vast reserves of water available.

And we generally assume that it's free, or almost free. But before clean water comes out of our taps, several things have to happen. We have to find a source of water, build machinery to pump it, piping to carry it, plants to treat it. Thanks to our treatment of water, chlorine has become an acquired taste in millions of households. We have to elect politicians who will run our municipal affairs, and look after our water treatment, and do the paperwork involved in supplying us with water. Once we get the water to our houses, we have to install pipes and valves and shut-offs and vents. We have to put in a separate line and a heater to heat some of the water.

Once we've got water, what do we do with it? We put it through our washing machines, toilets, sinks, dish-washers, car washes and pesticide-filled lawns. We use it to wash our windows, our sidewalks and streets. We spray it in the air for pretty fountains. We put out fires with it. We clean wounds with it. We make concrete with it. We use it in the production of plastics, steel and paper. We hose down chemical spills and industrial work sites with it. We clean paintbrushes in it. And we drink it.

What if we had water meters beside our kitchen sink? What if they read dollars and cents instead of gallons or liters?

Then we have to deal with getting rid of it. We need to build another whole network of drains to carry away our dirty water and sewage. We need to build treatment plants, and hire people to run them. And we need to elect politicians who will vow to "do something" to clean up the water that we've polluted.

The process costs billions of dollars worldwide, and still people suffer and die in many parts of the world for want of clean water, while we blithely open our taps and let our most precious resource pour down the drain.

There's not much we can do at home about the unequal distribution of water in the world. But the other major problems, contamination and waste, we can do something about. Although most of the advice in this chapter has to do with waste (we'll deal with contamination in other chapters), these two problems are connected in ways that might not be obvious.

The more we process our water, the more chance it has to become contaminated. That's because we have one sewage system for all purposes. We put our drinking water, our toilet waste and commercially contaminated waters all down the same system. We do our best to clean it up, then we pour it all out into the same river, lake or stream, and then we drink it again.

And of course, the more water we have to process, the more bleach we have to produce (which isn't a terrific thing to have around—it is, after all, a poison), and, naturally, the more we have to pay our governments for looking after all this stuff for us. So it's not so easy to keep cleaning our water.

Yes, we could be drinking Shakespeare's bathwater, but more to the point, will our great grandchildren be able to drink the water we used to hose down the dog? Will there be any clean water left?

Does it make any sense for us to save water at home? Isn't our home usage just a drop in the bucket, compared to what agriculture and industry use?

> **Household usage is about 5% to 10% of total fresh water used worldwide. Most of that is used in North America.**

On average each of us consumes nearly 53 gallons of water a day at home. Some citizens of water-poor countries survive on as little as 4 gallons a day. We've grown used to seeing water flow out of our taps and down the drains. What if we had an automatic shut-off on our household water that limited us to, say, 13 gallons of water a day?

What To Do

> **Turn the tap on briefly to wet your toothbrush, and turn it off until it's time to rinse.**

In our house, the average toothbrushing time is about a minute and 20 seconds. If we turn on the tap at the beginning of that time and don't turn it off until we're finished, we will have put down the drain approximately

2 gallons of water. In our little household of three people, we could waste over 4000 gallons of water per year just in toothbrushing.

Take the test in your household. How long does it take you to brush your teeth? Multiply that by the number of times you brush your teeth each day, then multiply that by the number of people in your household, and you'll soon see that you could have a terrific amount of water rushing uselessly down the drains.

My sister, Elizabeth, spent a great deal of time traveling the earth's oceans on sailboats, where she learned to brush her teeth with ¼ cup of water. The captain brushed without any water at all. We don't need to go that far, but we could all use less water than we do.

Keep a bowl or basin of water in the sink whenever you're working in the kitchen.

When I prepare meals or bake, my hands get sticky or greasy, or just in need of a quick clean every few minutes. You know that if you're peeling vegetables or coring fruit, or kneading dough, you have to keep washing your hands just so you can continue to work. But every time you turn on the tap, you run a terrific amount of water down the drain.

Usually by the time I need my first hand wash, I have at least one dirty bowl, so I stick it in the sink, run in some water, and leave it there. I keep the bar of soap handy. Then every time I need a quick hand rinse, I just swish my fingers clean in the bowl of water. When I'm finished baking, I add a little more hot water and soap, and wash the beaters or knives or mixing spoons right in the bowl. This saves a lot of turning the tap on and off, and a lot of water from going down the drain uselessly.

The bonus, of course, is that it makes your kitchen work easier and faster. My Dad says he can wash more dishes in less water than anyone in the world. He claims he can do a sinkful of dishes in a teacup.

Keep a bottle of water in the fridge.

We use bottled water—from the tap. Have you ever let the tap run for a minute to get an ice cold drink? About 15 years ago, I filled an empty soft drink bottle with tap water and stuck it in the fridge. That same bottle is still in our fridge today. Of course it has different water in it.

Our water bottle has its own spot, in one of those bottle hangers that goes under the fridge shelf (it's been in the same place for years, even when we've moved houses and changed fridges), so that we never have to run the tap for a drink of water. It's always cold and handy. If you're just starting this system, be sure to label the bottle "Drinking Water." Once, years ago, when my Dad was visiting, he took a big swig from the bottle in our fridge, only to discover that someone had put a bottle of white rum in to cool.

Take a five minute shower instead of bathing.

Abandon the bathtub, and hit the showers. Sometimes it just feels great to soak in the tub, but that tub holds between nine and 33 gallons (40 to 150 liters) of water, depending on how full we fill it. We'd have to shower for 15 minutes before we used up the quantity of water it takes to fill the tub. When we were kids, we used to share a bath. We thought it was fun, but little did we know that our smart parents were saving on water heating. My daughter Caroline still enjoys a bath with her little cousin, Lisa.

Turn off the shower while you lather your hair and body.

My sister Elizabeth, now home from the seas, lives on the top floor of a house where the water flow is often restricted by the people showering on the floors below. Subjected to bursts of either scalding or freezing water in the shower, she learned this trick out of self-preservation. She wets herself down in the shower, turns off the water, lathers her body with soap, her hair with shampoo, and then turns the water back on again to rinse. And of course, we all remember the other popular way to save water: shower with a friend.

Learn the cold water hand wash.

If every time you wash your hands, you turn on the hot tap and wait for the water to get warm, you could run anywhere from a few cups to a gallon or more of water down the drain. There are two problems with that.

First, it's water that has gone through the entire system of our waterworks for nothing. It's been pumped from the lake or river, using energy, it's been bleached, it's been pushed through miles of pipes, and then it just goes back down the drain to be processed all over again with our sewage, having done nothing.

Second, it's water that's already been heated in your home water heater, but has cooled before it gets to you.

The energy that was used to heat it, which you pay for, has been wasted.

I even wash my face in cold water every morning and night. I'm trying to convince myself that cold water is kinder to my skin than hot, but frankly, I know of no studies that would back me up on this one. However, my partner, Barry, tells me he once read that Paul Newman soaks his face in ice water to stay young looking, so maybe I'm on to something here. Masochistic as it may sound, I find it refreshing to start my day with a cold splash. I confess that so far I've made very few converts to this theory, but I still swear by it.

Do you get as clean with cold water as with warm? The answer is yes, although there are exceptions. If your hands are greasy or oily, warm water will help to dissolve the grease or oil more quickly than cold water. But for ordinary, garden-variety dirt or stickiness, cold water works just as well as warm.

What about germs? Ordinary hand soap will take care of whatever germs are washable. If you wanted to be totally antiseptic, you would have to use boiling water, probably for several minutes. I think most of us would opt for just plain clean, thanks anyway.

> **Fill a large plastic yogurt or cottage cheese container with water, and put on the lid. Then slide it down into the space in the toilet tank behind the flush handle.**

The idea, of course, is to keep the water level in the tank high, while using less water. Traditional North American toilets use about four and one-half to five gallons of water per flush.

In the sixties, when we first started becoming environmentally conscious, we used to put bricks in our toilet tanks. Some households still do that. The brick in the tank simply displaces some of the water used on each flush so that less clean water goes down the drain. Use a very clean, solid brick (new, if possible, or one with no mortar clinging to it), or wrap it tightly in a plastic bag so that no particles will float out and clog your toilet.

We use the plastic container method in our house. If you can't fit one in your particular brand of toilet, try an empty mayonnaise jar, a margarine tub, or a soft drink bottle with a screw-on lid. This water-saving idea takes literally two minutes or less.

You can also buy a toilet dam, which sticks on the inside of the tank. But in the meantime, why not just take two minutes and try the quick and easy home remedy?

Fill your coffee pot with water the night before.

We've been told lately that we have to run our taps for several minutes in the morning to get rid of the lead that may have accumulated overnight from the solder in our plumbing pipes.

Of course that's not necessary if you do all your showering and flushing before putting on the coffee. All that water usage in the morning will get rid of the lead in the pipes by the time you want to fill your coffee pot. But if you're the type of household that stumbles from the bedroom and turns on the coffee first, and you've filled the pot the night before, you won't have to run water down the drain uselessly.

If you have an automatic coffeemaker, you already

know how nice it is to come out in the morning and just flip the switch, when the water and coffee are already in the machine. Now you can give yourself a gold star for saving water by setting it up the night before.

Put a pail or a wash basin or watering can, or even a bowl or a coffeepot, underneath a dripping tap.

A dripping tap can lose anywhere from six to 20 gallons of water per day. It's not always practical to run to the hardware store the second the tap starts to drip and get a new washer and run home and fix it. In my experience, the tap usually starts to drip just when I've changed into my best clothes and I'm rushing off to an important appointment. But until you can get at the repair, at least don't waste the water.

Use the saved water on your flowerbeds or houseplants. Wash your underwear in it. Put it in the dog's water bowl. Make coffee with it. Wash the kitchen floor with it. Wash your fruits and vegetables in it. Wash the car with it. Don't let it just run down the drain. Then fix that tap as soon as possible.

Turn off the shut-off valve on a leaky tap for the time you'll be away.

The United Nations estimates that as much as one half of the total water supply of cities is wasted through leaks in pipes and taps. If you're going away for vacation, or even for a weekend or a day, it only takes a second to turn off the valve. You'll have peace of mind while you're away, knowing that the leak hasn't suddenly become a flood, and drowned the cat while you're gone.

Each time you want to wash the car, put it off for one day.

We like our cars clean and shiny. And in fact, up to a point, washing your car can help to preserve it. But only up to a point. Car washing consumes gallons of drinkable water with every washing. How often do you wash your car? Could you put it off for one day each time? Let's say you wash your car once a week. That's 52 times a year. If you put it off one day each time, you'll be washing it only every eight days. That's about 46 times a year. So already you've saved six washings—and 60 to 120 gallons of water.

A large portion of our car wash water just runs down the street without ever touching the car. That's because we often put the hose on the ground and let it run while we rub down the car. There's a really quick solution to that one: shut off the flow. In lieu of a hose attachment, you can simply put a kink in the hose to temporarily halt the flow of water, although that can eventually wear a thin spot on the hose.

Use a bucket instead of a hose to wash the car.

A couple of gallons of water in a bucket will wash the car just as well as a running hose, with a lot less waste of water. But when it's time to rinse, a quick spray with the hose will likely use less water than a bucket rinse.

Revive the rain barrel.

Many countries still use them, or variations on them. Bermuda has almost no fresh water, and relies on rainfall for domestic consumption. A water collection tank on the roof is a common sight.

We may not have to resort to those measures, but a rain catcher in the back yard makes a lot of sense. If you can find an old wooden barrel, you can make it a picturesque part of your landscaping, but even that's not essential. A large plastic garbage can will hold 15 to 25 gallons of rainwater. When it's time to wash the car or water the plants, just dip the bucket or watering can.

Mom and Dad have a rain barrel out by their garage, and it's very handy for watering their hanging outdoor plants. No need to haul out the hose—just dip the watering can. I always remember Mom collecting rainwater to use in her steam iron, so the minerals in the tap water couldn't clog up the iron. Nowadays, you can buy "demineralizers" for your steam iron water, but if you have a rain barrel, that becomes an unnecessary expense.

And remember that the water from your rain barrel hasn't been chlorinated, fluoridated, pumped and priced (at least not this time around). In the city, your rainwater may be dirty with atmospheric pollutants, but your plants will probably still like it better than heavily chlorinated water, and your car certainly won't care.

Only do the laundry when the washer can be filled.

Do you have a full washer load? Are you sorting unnecessarily? Do you really need to do your whites separately from your colors? What if you put them together to make a whole load instead of two half-loads? I do it all the time, and my clothes still come out just fine. In fact, the only time I sort clothes is when I have enough to do two or three loads on one laundry day. Of course if you do your laundry at a laundromat, you already know the savings involved in doing one load instead of two.

If you have to do laundry before you have a full load, set your water level control for a smaller load.

If your machine doesn't have a control, you can simply turn the dial to wash when the machine is partly full of water. Reduce the amount of laundry powder proportionately—to find out why, see Chapter 3 on home cleaning.

Check the weather forecast before watering the lawn. If you do sprinkle the lawn, write a note that says "Turn off sprinkler," and leave it sitting out on your counter until you actually do turn the sprinkler off.

Lawns are a real luxury, and one that we might seriously consider trading in for a better idea. But in the meantime, have you ever seen lawn sprinklers going full tilt during a rainstorm? Use the fresh free stuff that's pouring out of the sky.

You might want to put a time on your reminder so if someone else notices the note, they'll know when to turn off the outdoor faucet. If they don't go and do it themselves, they'll certainly remind you to do it.

> *Think of saving water this way: what if you had to carry home all the water you needed every day—in jars on your head?*

Chapter

2

SAVING PAPER

Why Save Paper?

Whenever I've flown over our continent, I've looked at the endless stretches of green and thought, "There are enough trees to last us forever."

But of course, there aren't.

A single edition of the Sunday *New York Times* newspaper uses 75,000 trees. What about Monday, Tuesday, Wednesday...? What about the *Washington Post*, the Chicago *Tribune*, the Toronto *Globe and Mail*, the *Los Angeles Times*, the *Miami Herald*?

All that paper comes from our forests. The endless stretches of green may not be so endless after all. But we've been led to believe that every tree that's cut is replaced.

Canada, which supplies almost all the world's newsprint, cuts down 247,000

acres (100,000 hectares) per year more
trees than it replants. The United States
loses an acre of forest every five seconds.

My local government has admitted that only 26.5% of
the forest near an area that environmentalists are strug-
gling to protect was replanted in the 1980s. Some areas
are never replanted. Improper logging practices have destroyed any possibility of regenerating forests where they
once stood. The loss of forests soon leads to soil erosion
and flooding. New trees cannot gain a foothold. The area
is simply written off, and left a shambles.

But even replanting doesn't give us the forest we originally had. Replanted forests often resemble columns of
soldiers lined up in orderly rows. No bushes or wayward
trees mar the clean lines from one end to the other. No
plants crowd into the spaces between the rows. We don't
see deer or bears living in that kind of a forest. In an old-
growth forest, numerous species of plants and animals
live in the diverse "understory." Their existence depends
on this growth of bushes, flowers, ferns and smaller trees
that live beneath the canopy of the oldest trees. When a
forest is clear cut, the understory disappears along with
the desired trees. The habitat for the forest species is lost.

If the earth heats up much more, because of the greenhouse effect, some of our replanted areas will not survive
as forests. Many of the immature replanted areas are
already showing signs of ill health. The young seedlings
may never become mature trees.

We can't blame all of this deforestation on our insatiable
need for paper. Agriculture, urbanization, recreation and
industry all take large tracts of our forests—but we are,

as individuals, prolific and wasteful paper users, both at home and in the office.

Some 40% to 50% of what we throw away is paper.

Think of any office—all those filing cabinets full of paper, all those boxes of paper by the copying machine (and all those garbage bins of spoiled copies), all those reports, all those paysheets, checks, applications, submissions, letters, envelopes, computer paper, phone message slips, file folders, and all those duplicates, triplicates and quadruplicates of information useful and useless. Then picture that one office multiplied by all the offices on the same floor of the building, then multiplied by the number of floors in the building, then multiplied by the number of buildings in the city or town, then multiplied by the number of cities in the country, on the continent, in the world.

It takes 17 trees to make one ton of paper. Depending on what newspapers you read, you could be using up one tree every 10 or 12 weeks—week after week after week. . . .

I wonder how many trees ordinary individuals could save in a year with home and work conservation practices?

Also bear in mind that paper production is extremely hard on our water systems. We bleach wood pulp to make white paper products, including everything from office stationery, computer paper and books, to diapers, sanitary supplies, milk cartons and coffee filters. In the process, numerous toxic waste products, such as deadly dioxins and furans, are created and dumped into our water systems. Neither is recycling always the answer. De-inking, bleaching and processing used paper causes similar problems.

Nearly half of all untreated effluents discharged into our water comes from the pulp and paper industry.

You can't prevent all this by yourself. But simple conservation practices can—and will—help. Furthermore, just like most conservation methods, you'll likely save money while you're at it.

What To Do

For telephone message pads, grocery lists or notes for the bulletin board, use the back side of flyers or notices.

Most people keep a pad of paper by the phone for messages or in the kitchen for grocery lists, or on the bulletin board for notes. Why would anyone buy scrap paper when they get it delivered free to their door every day?

Here's how we get scrap paper at our house. During the past week, I received three notices from my child's school, a flyer from a roofer, two flyers from the community center, an ad from a real estate company and an ad for a garden tour. Each of these notices was printed on one side only of colorful 8½″ × 11″ paper. As each notice arrived in the house, I threw it into a file folder that I keep in the kitchen drawer.

When I have a small fistful of these pages, I cut the whole stack into quarters, giving me a nice little pile of multicolored scrap paper. Sometimes I staple a bunch together to make a pad, but you can keep them loose or in a little box. By the time I need more, I've usually received more flyers.

Keep a scrap paper box beside the copying machine.

Do you work in an office? Is there a waste basket beside the office copier? All those hundreds of ruined copies, still good on one side, represent tons of paper. I bet you can find a cardboard box that once contained letter- or legal-sized paper. Take one minute to write a little sign to post above the copy machine. Encourage your fellow workers to use that source for scrap paper, rather than always reaching for a new sheet.

You can even grab a fistful of this "good one side" paper, cut it into quarters and staple together your own phone message pads or doodle pads. If you take messages for someone else, impress them with the fact that you're saving the company money by reusing paper that would otherwise clutter up our landfill sites.

My publisher can afford all the note pads they want, but started to follow this advice after I delivered the first draft of this book to my editor.

Write non-business letters on the back of scrap paper.

Your correspondents may be impressed enough to emulate your example, and you might receive return letters on some interesting paper. Once my cousin and I sent the same Valentine back and forth to each other for about six years, adding a new message each time.

Since I've been computerized, I often print out letters on the back side of rough draft or ruined computer paper. If you don't tear off the punched edges of your computer paper, you can easily turn it over and print on the back.

During the U.S. Civil War, paper was scarce. Soldiers "turned" letters, and wrote across them. After the Americans left Vietnam, families sold old letters and receipts because the paper was so valuable. Nothing was written on one side only.

We may not necessarily go that far, but in saving paper, thrift becomes a virtue. If we don't waste paper, we won't be wondering in five years where our next tree will be coming from.

Don't throw perfectly good envelopes in the garbage just because they have an unwanted address on them.

Does your trash can get filled with empty envelopes? I pay most of my bills at the bank, as do many people. But the folks who send out the bills usually send you envelopes along with their demands for money. I just use a black marker to cross out their address, and write my correspondent's address in the rest of the space. Sometimes I use a gummed label over the address.

If the envelopes already have the postal or zip code imprinted in bars, be sure to cover the bars with your label so you don't incur the wrath of the post office—or risk having your mail go astray—with confusing signals. Amuse and delight your correspondents with the variety of different envelopes in which they receive your wonderful letters.

You can even reuse envelopes which have been addressed to you, by opening them carefully, labeling over the address, and resealing them with sticky tape. But please don't try to cheat the post office out of the postage by reusing the stamps—they need the money!

My publisher gave up its corporate ego along with its corporate logo when it began reusing envelopes. They had a special stamp made (in green ink) that reads: "We reused this envelope to help protect the environment. We hope you can do the same."

Reuse greeting cards at least once.

Christmas cards, birthday cards and all other kinds of greeting cards can usually be used more than once. Sometimes you can cut smaller cards from larger cards. Or you can often cut off the right hand side of an open card, where the personal message is penned, and use the front as a gift tag, writing on either the front or back of it. My sister-in-law Marlien uses pinking shears to cut out the cards in fancy shapes.

My mother-in-law sometimes sends cards with a letter

inside, and doesn't sign the card. I'm never sure if she's doing it to save the card, or if she just forgets, but I always appreciate it. Of course, the best cards are homemade ones. I often cut out odd little shapes out of the blank parts of flyers and decorate them with colored pencils or stickers or photographs. Then I write my own messages inside.

Reuse gift boxes as often as possible.

Save those shirt boxes! We have a corner of a closet that's devoted to empty boxes. Whenever we receive something in a box, we stash the box in the closet. So when I go to the store to buy a present and they ask me if I want a gift box, I can say, "No thanks, I already have one." Some of the boxes in our family have made the rounds from sister to brother to parents to nieces and nephews dozens of times.

Use a rag instead of paper towels.

Picture for a moment the mountains of paper towels in every supermarket in every city in the country. These were all made from our forests, by cutting down trees, transporting them, mashing them into wood pulp, bleaching them, packaging them, all of which uses energy and is destructive to our environment. Then picture that same mountain of paper towels, all scrunched up with little blobs of spilled milk or jam in them, filling our landfill sites. Talk about waste.

I'm appalled when I see people reach for a paper towel to wipe up a teaspoon of spilled grease. Keep a rag under the sink just for spills.

About once a week when I was a child, my Dad would say, "Everything that comes into this house, I have to carry out again as garbage." A one-use paper towel certainly creates garbage in a hurry.

When my dishcloth gets too disgusting for dishes, I downgrade it to a floor rag. When it's too worn for even the floor, it goes to the basement or the garage.

Use a permanent or reusable coffee filter.

These are generally conical, like your plastic filter holder. They're made from very fine metal mesh, and are available at many kitchen stores. Another option is to buy or make an unbleached cotton filter to reuse.

The bonus is you get no dioxins in your coffee or your compost. The dioxins that end up in our water from producing bleached paper can end up in the paper products themselves. If you throw paper coffee filters into the compost, you're liable to put dioxins in your soil.

If the thought of dioxins doesn't bother you, you can reuse your paper filter several times. Just dump the coffee grounds into your compost, and leave the filter in its plastic holder while you rinse it out.

Use cloth diapers instead of disposable plastic and paper diapers. Encourage your daycare or nursery school to do the same.

Cloth diapers today bear little resemblance to those of just a decade ago. Look for fitted diapers with Velcro closings, and waterproof diaper covers that hold the diaper without pins. Compare a plastic diaper next to your cheek with a cotton one. There's no comparison.

At seven ounces per used disposable diaper, an average baby will account for over 3000 pounds of waste in a landfill site over a period of two and a half years. One billion trees per year go into the manufacture of disposable diapers.

Disposable users and manufacturers often tout their convenience. But how convenient are they? People who use them have to buy diapers whenever they go shopping. They likely make many special trips to the store because they're out of diapers. With cloth diapers, they would spend about 10 minutes per laundry load—30 minutes a week—for the two and a half years the child is in diapers. One extra trip to the store takes that long.

Then there's cost. Studies show disposables cost most, a diaper service less, and laundering your own the least. How many hours do you work to pay for disposables?

For every dollar spent on disposables, the taxpayer spends 8% more to dispose of them. This doesn't account for the environmental costs, which are not yet calculated.

Disposable users are supposed to dump the poop in the toilet. But many just wrap it up and throw it away. Our landfill sites are not as equipped to handle raw human waste as our sewage systems are.

So-called "biodegradable disposables" are made with a starch filler. These plastics need either sunlight or soil contact to break down. In a garbage dump, few diapers will get either. Even if the starches in them do break down, we are left with many tiny pieces of plastic, instead of one large piece. They still last forever.

There are alternatives. In many cities and towns, diaper services are springing back to life as more and more people discover the benefits of beautifully washed, picked up and delivered cloth diapers.

Every week you put your dirty diaper pail outside your door, and a bag freshly laundered diapers comes back. You never have to carry disposables home from the store. You never carry mountains of garbage out to the curb, full of one-use diapers.

Not everyone is lucky enough to have diaper service available. If you launder your own instead of using disposables, you could save as much as $2000 over the diapering years, enough to pay for a washer and dryer.

> *Paper surrounds us. We use it dozens, maybe hundreds, of times every day without thinking. But saving paper is one of the easiest and most beneficial contributions we can make to our environment.*

3

NON-TOXIC CLEANING

Your Home, Your Body And Your Laundry

Sometimes, when I think of all the guck we dump into our lakes and rivers, I hate the taste of tap water. When I think of all the bleach, lead, dioxins, heavy metals and other pollutants that could be in our municipal water supplies, I'm tempted to get a home water purification system or buy only bottled water. What does our drinking water have to do with cleaning our homes, our laundry and our bodies? Simply this: most of the cleaning materials that we use at home end up going down the drain, into our sewage systems, and back into the lakes, rivers and streams from which we take our drinking water.

What's worse, in some places the sewage goes untreated, directly into the water. Halifax, on the Atlantic coast, has an enormous problem on its hands, trying to figure out how to clean up a three-foot-deep layer of filth on its harbor floor. Victoria, north of Seattle on Canada's

Pacific coast, also dumps its sewage untreated into the ocean. While politicians and environmentalists argue about what should be done, some sensible folks are suggesting reducing the amount of sewage and other contaminants before it ever gets to the harbor.

But Halifax and Victoria are not alone. Many cities on the coasts of North America do the same. According to the United Nations, most of the world's wastes—20 billion tons of it a year—end up in the sea, often without any preliminary processing.

Despite the abuse of our water supply, I still drink the water from my tap for several reasons.

First, in developed countries, there are no guarantees that any purification system or any bottled water is better for you than your locally treated tap water. Independent studies have shown just the opposite to be true in some cases. Many places do not have any regulations governing bottled water or purification systems. You have to rely on the manufacturer for information about any particular product. Of course, they won't tell you that your tap water is just as good for you. Why should they?

> **Fine print disclaimer in advertisement for a water filter system: "The list of substances which the treatment device removes does not necessarily mean that those substances are present in your tap water."**

Second, even if I could be sure that I was getting superior water, I'd just be perpetuating the notion that good water is for those who can afford it. What if my neighbor can't? Who would decide who should have good water? Do big corporations get good drinking water because they can afford it, while subsidized housing projects drink the other stuff? Should hospitals and schools get it? Where do you draw the line?

Third, using bottled water or a home purification system would relieve me of any responsibility to ensure that our lakes, rivers and streams are cleaned up. The contamination of our waters would become "their" problem instead of "our" problem. If I'm sure that my water is safe, why should I worry about anyone else?

That kind of attitude indicates a bunker mentality. What if my place of business or my child's school uses only tap water? I don't relish the thoughts of walking around with a water canteen slung over my shoulder or sending my child to school with a bottle of water. Shouldn't clean water be available everywhere? If I want to maintain the freedom to live, work and play wherever I choose, then I should be able to rely on the water I drink, wherever I go. If it's so polluted that I can't drink it, would I want to swim in it or fish in it?

Fourth, much of the bottled "spring water" or "mineral water" comes from under the ground or from glacial water sources. Using this supply of water is not the same as using the ever-recycling water from our lakes and rivers. "Ground water" or "fossil water" accumulated hundreds of thousands of years ago, over a period of thousands of years. All over the world, the depletion of our ground water and glacial water sources continues at an ever increasing rate. When it's gone, it's gone. The wells will simply run dry.

That means sinking lands, sinking water tables, and

possible loss of habitation for some species. We're not talking hundreds or thousands of years. Some of the large underground water supplies are not expected to last another 50 years. If consumption continues at the present rate, these sources could be gone within my lifetime.

Fifth, the energy needed to manufacture, transport, and dispose of the packaging for bottled water is enormous. The manufacture of both glass bottles and plastic containers requires heat, electricity and water in great quantity. The burning of fuel to provide the heat and electricity produces pollution. The diversion of water from more beneficial uses is wasteful. We burn still more fuel transporting and distributing this packaged water. Finally, we must dispose of the water container in one way or another, either by reusing it, recycling it, or dumping it in a landfill site. Each of those methods wastes still more energy.

I occasionally buy mineral water for a "treat" (although I guess I've really just talked myself out of it), but I would hate to have to rely on bottled water for my everyday requirements.

We can make a start on cleaning up our water, whether or not our waste water gets "purified" before it gets recycled. We read every day, to our dismay, of many different toxic substances found in many bodies of water. At home we can help keep our water clean by not putting pollutants down the drain that might never leave the water system—pollutants that we don't want to be drinking.

North America has one of the most valuable and vulnerable resources in the world: the Great Lakes. Because they are so huge—along with the St. Lawrence River, they are largest surface expanse of fresh water in the world—we have treated them as if they would continually provide us with clean water. We were wrong.

Only 1% of Great Lakes water flows out of the Great Lakes basin every year via the St. Lawrence River. Whatever we put into the lakes stays there.

When I first came to live in Toronto, fresh from the Canadian prairies, I reveled in the joy of swimming in Lake Ontario. Now, less than 20 years later, I find our public beaches posted with signs warning against swimming in the polluted water. I take my daughter to Sunnyside Beach, on the shores of that beautiful Great Lake, where she can only swim in a swimming pool.

Many groups and individuals are working to rid the Great Lakes of contamination, and I support them wholeheartedly. If you want to help support them, see the chapter on environmental groups. But there's plenty that you and I can do at home to help keep everyone's local water supply cleaner.

"A pool supply company was fined $4000 after spilling chlorine that killed hundreds of fish."—news item, 1989

Is whiter than white a virtue? One of the strange perversities of human behavior is that each time somebody comes up with a way to make cleaning easier, we raise our standards of cleanliness. Before vacuum cleaners were invented, you can be sure that a lot more dust bunnies were allowed to congregate under beds and behind dressers than nowadays. Once a year "spring clean-ing" took care of accumulated dirt. Now we vacuum weekly.

Before washing machines were invented, clothes had to be dirty, not just worn once, before they were laun-dered. Now, since it's easier to launder, we launder once or twice or three times a week, often throwing clothes in the wash simply because they've been worn, not because they're dirty.

When I was growing up, our household operated on the "clean because company's coming" standard. Don't think we were slobs. In fact, our house was pretty clean. We did dishes, swept the floors after every meal and we always wore clean clothes. We tidied up fairly regularly. In fact, my mother worked pretty hard at those thankless tasks.

But she had more important things to do than scrub every little corner of the house. (Dad too, but women were in charge of the domestic chores in those days.) Mom was always available to her four busy kids to help with projects, school plays, puppet making and fort building. And the nifty bonus benefit was that everybody liked coming to our house. All our friends liked my mother, and nobody ever complained about less-than-hygienically sterile bathroom sinks. Nobody ever suggested that my mother ought to be scrubbing out the bathtub instead of helping with our dried leaf collection or adjudicating the school drama festival.

While I realize that nobody wants to live in filth, we can start by taking a look at whether we really need to clean as much and as often as we do. Let's try to keep a perspective on what's really important in life. Will our friends like us less if our window ledges aren't washed? Will they talk behind our backs if they can't see their reflection in our kitchen sinks? What kind of friends do we want anyway—sanitary inspectors?

What To Do

> **Take a a few extra steps down the super-market aisle and pick up a box of laundry soap instead of detergent for your normal washing.**

Laundry soap is much more benign than detergent. For everyday laundry, it works just as well. I've been using it for several years, interspersed with an occasional use of detergents.

Try it once and you'll see how easy it is. Most of us are not car mechanics or coal miners. We don't get our clothes so filthy that we need that heavy-duty cleaning power promised by detergent manufacturers.

Day after day, week after week, the cleaner we use the most at home is laundry detergent, so naturally that's a big factor in our concern for water quality. What's wrong with it? Most detergents are petroleum based. Therefore, when they go down the drain and out into the waste stream, they break down very slowly, if at all, sometimes leaving residues that don't disappear.

Phosphates, which help to soften the water in washing machines, are another big culprit found in most detergents. Phosphates fertilize the natural algae in our lakes, causing them to grow at an explosive rate. This excessive growth of algae rapidly uses up the oxygen in the water, choking out other marine life, such as fish, shellfish, and smaller aquatic animals and microorganisms. That's what happened to Lake Erie 20 years ago. Remember when they said it was "dead"? Luckily, that pronouncement was premature. The United States and Canada got together, and, among other things, restricted the amount of phosphates allowable in laundry detergents. (Unfortunately, they did not apply the same regulations to automatic dishwasher detergents, which are a growing problem.) Today Lake Erie, although not cured of all its ills, is on the road to recovery. However, even if the percentage of phosphates is reduced in laundry detergents, as the population continues to grow, the total amount of phosphates used will increase.

Other things which may be in your detergent include enzymes, perfumes, optical bleaches, chemical bleaches and chemicals that bind heavy metals in water, all of which can be harmful to the water supply. As bleach breaks down in the environment, many toxic and carcinogenic substances are produced.

As for other home cleaning products, the list of ingredients sounds like a deranged chemist's shopping list. Drain cleaners usually contain lye, which is both caustic and poisonous. We could also be pouring down our drains sodium hydroxide (caustic soda) or potassium hydroxide (caustic potash), sodium hypochlorite or hydrochloric acid (muriatic acid). Sound good so far?

Toilet cleaners may contain hydrochloric acid, oxalic acid (a poisonous compound), paradichlorobenzene (widely used as an insecticide in mothballs—when was the last time you saw moths in your toilet?), or calcium hypochlorite.

Silver cleaners can contain sulfuric acid (also called vitriol, from which we get "vitriolic"—corrosive, burning or caustic). Disinfectants may contain phenols (carbolic acid), and rug cleaners may contain naphthalene (also used in mothballs).

I'm not a chemist. I don't know what every one of these chemicals does. But I do know that most of this stuff doesn't sound like anything I'd like to have in my drinking water. Luckily, there are a lot of options available.

Next time you're in the supermarket, pick up a box of soap flakes instead of detergent. Unfortunately, your choice may be severely limited, and you may have to look carefully. In my supermarket there are dozens of varieties of detergents, and only one brand of laundry soap, Ivory Snow. Nevertheless, you should be able to find at least one type, often with a baby's picture on it, because laundry soap is seen as being non-irritating for baby's skin—another clue to its environmental friendliness.

Give it the home laundry test. Do a couple of loads of laundry using soap instead of your regular detergent. You'll likely find, as I did, that you can't tell the difference in how clean your clothes come out. If you have very hard water, or think you need a fabric softener, use ½ cup of vinegar in the rinse. Do not mix chlorine bleach with vinegar. You could end up with toxic vapors.

Other environment-friendly laundry powders are starting to appear on the store shelves. You might want to try a cleaner called Ecover, which is endorsed by Friends of the Earth. When I have laundry that needs a little more power than soap, I've used it with success. Tide Free has no phosphates and no perfumes. Echo-Logic distributes a concentrated laundry detergent called Laundry Miracle, with no phosphates, enzymes or nitrates. Call (416) 360-8799 for mail order. Check out your health food store for other safe cleaners.

Use borax or washing soda instead of bleach.

I would only use chlorine bleach in the rarest of circumstances, for otherwise untreatable stains. I certainly don't recommend it as a regular additive to your laundry. That powerful bleaching action continues even after the bleach goes down the drain, and can cause havoc in our water systems.

If you really think that laundry soap needs a bit of a boost, try a little borax or washing soda, both of which can be found on the shelves near the detergents. Either of them will do a good job of boosting your soap power by softening the water, while being much less harmful than bleach to the environment. They are also gentler on your clothes.

Borax, however, is an ingredient found in many fertilizers, and although it does inhibit molds, I fear that great quantities may cause problems similar to those caused by excessive use of phosphates. I use a little washing soda now and then, but it can cause irritation on your skin, especially if you handle it when your hands are wet. Either of these products is preferable to bleach.

Buy a plunger.

Drain cleaners are so toxic and corrosive that they should never be discarded with your regular garbage. If your community does not yet have a hazardous waste disposal program, store drain cleaners in a safe place until you do have such a program. (Soon all communities will have to face the music.) For now, arm yourself with a plunger, also known as a "plumber's helper." It's a good idea to have one for emergencies. A few forceful "plunges" will clear almost all blocked toilets.

If you've got serious underground drain problems, you'll need a plumber's "snake," in which case the drain cleaners won't do much good anyway. You can buy a plunger at the hardware store, and you can rent a snake at any tool rental company. (If you live in an apartment, get your landlord to call a plumber for the snaking operation.)

Clean your drain.

If your drain is starting to slow down, do some preventive maintenance before you need the mechanical aids. Pour hot water down the drain, then add ½ cup of washing soda. After 15 minutes, pour more hot water down the drain. If you've been using washing soda in your laundry, as suggested above, it will help keep your laundry drains clear.

Here's another recipe for slow drains. Pour in boiling water. Add ½ cup of baking soda, ½ cup white vinegar and ⅛ cup of salt. You shouldn't have to resort to this if you use the quick and easy sink cleaning recipe below.

Your sink drain will get a regular dose of baking soda and vinegar, which will help to keep it clear. No need to lock these environment-friendly cleaners away in a safe place. We keep them on the kitchen shelves with our groceries.

Stop buying special toilet cleaner.

Do we really need to clean our toilets with every single flush? By the time you've installed one of those hangers in your toilet tank, you could have cleaned the bowl. Sprinkle a little baking soda on your toilet brush, swish it around the bowl. That's it. Nothing more required. If you feel that it's necessary to disinfect toilets, pour ½ cup borax in the bowl, and let it stand for an hour. Nothing toxic, nothing corrosive.

Clean sinks and appliances with baking soda.

Baking soda is a terrific cleaner that is mild on the environment. You can get your kids involved in cleaning by showing them this nifty and reliable little science experiment. Pour ½ cup vinegar in your sink (with the plug in), swish it all around the bottom and sides, and then let them sprinkle on some ordinary baking soda. Surprise—the resulting fizz is very gratifying. Let it sit for a minute or two, then wipe it clean with a rag. After cleaning sinks with vinegar and soda, you can throw a little salt and hot water down the drain if necessary to help keep the drain free running.

I use the same combination of vinegar and baking soda

on countertops, appliances and for general cleaning. If you just want to wipe up the sink quickly, you don't even have to use the vinegar. Baking soda works well all by itself. Besides the soda that I use in baking, I have another box of baking soda that serves double duty. I keep it open in the fridge (just like your mother used to do) to absorb odors. But when I want to do a quick sink clean, I open the fridge door and grab the soda. That uses up the top layer of soda, exposing fresh soda to absorb more odors in the fridge.

If that's not convenient, you can just put the box from the fridge under the sink to use as a cleaner when it's outlived its usefulness in the fridge. As you replace each box in the fridge, downgrade it to cleaner status. Baking soda works just as well as a cleaner whether it's old or fresh.

Mix your own all purpose cleaner.

What about floors, and heavier cleaning? What about those days when you feel really energetic, and want to wash down the walls and clean the whole house? Here's a simple recipe for an all-purpose cleaner you can use without guilt.

First, a caution. *Do not mix this cleaner with bleach.* Any cleaner with ammonia in it will make deadly fumes when mixed with bleach. Ammonia is a good cleaner which is both cheap and biodegradable, two good reasons for using it instead of commercial cleaners. But it is corrosive and the vapors may be harmful. For that reason, I limit my use of it to heavy duty cleaning. You can usually find ammonia with the cleaners in a supermarket or hardware store.

The recipe for an all-purpose cleaner: Put about ½ gallon of water in a pail. Add ½ cup of ammonia, ½ cup of vinegar and ¼ cup of baking soda. That's it. It takes less than two minutes to mix it up. Then just get ready to scrub. This cleaner works well on floors, walls, bathroom sinks, appliances, counters, etc. Don't use it on copper or aluminum—including your aluminum kitchen sink. It's a good idea to have a little ventilation when working with ammonia.

> **Put a small bowl of ammonia in your dirty oven instead of more harmful oven cleaners.**

I'm not a big fan of oven cleaning. My sister and I like to joke that it's time to sell the house or change apartments when the oven gets dirty. If your oven is really dirty, try this trick after you've used the oven some evening, and it's still warm. Pour about ¼ cup of ammonia in a shallow bowl (like a soup bowl), and fill it up with warm water. Then put it in the oven and close the door.

At this point, I make a sign to tape to the oven door: "Ammonia in oven!" as a reminder not to turn the oven on. The next morning, open the windows, then open the oven door. Let the fumes out for a few minutes, then wipe down the inside with a damp rag. Again, be sure not to let the ammonia come into contact with copper or aluminum (watch out for your oven thermostat).

Substitute some quick home chemistry for specialized cleaners that contain toxins.

Here are a few useful cleaning recipes:

• Furniture polish: Mix ¼ cup of vegetable oil (any cheap cooking oil will do) and ⅛ cup lemon juice. Rub on furniture, and polish with a soft cloth. I suppose one could use any leftovers on a salad? For a non-edible furniture polish, simply use a damp cloth dipped in plain mineral oil (although it wouldn't poison you if you did happen to ingest some).

• Silver polish: Here's another nifty science experiment you can do with the kids. There are a couple of ways you can do this—if you have an old aluminum saucepan that you don't mind getting dirty, go ahead and use it. Otherwise, put a piece of aluminum foil in the bottom of an enamel or stainless steel saucepan. A wrinkly, used piece of foil works just fine. Pour in about a quart of water. Add one tablespoon of baking soda and one tablespoon of salt. Boil this mixture, then add your silverware. Like magic, the tarnish leaves your silverware and coats the aluminum foil. Get the kids to polish the silver dry while they try to figure out the chemistry. Some people warn of dulling the finish, but I haven't found that to be a problem. Another option is toothpaste and toothbrush. Use an old toothbrush with a little toothpaste on it to scrub those curly crevices in your silverware. Rinse in hot water and polish dry.

• Copper cleaner: Make a paste of equal amounts of salt, flour and vinegar (the flour just helps it stick together). Rub it on the copper items, and let it sit for 10 minutes. Shine clean with a soft cloth.

Use cedar to repel moths, instead of moth-balls, which contain poisons.

Cedar chips or cedar lining in a closet will repel moths. Look for cedar sachets, cedar balls, even cedar clothes hangers in hardware stores, sporting goods stores or department stores. You can use a woodplane to make cedar shavings from cedar wood or shingles that you may have on hand. Hang a net bag full of the shavings in the back of the closet.

Choose an environment-friendly automatic dishwasher powder.

Check your supermarket for new automatic dishwasher powder, specifically marketed as being environment-friendly. Some are now available without phosphates. The Soap Factory's Dish-a-Matic, without phosphates or enzymes, is available by mail order from Echo-Logic (416) 360-8799. Try cutting back on the quantity of any cleaner you use (see end of chapter).

Try plain water first.

Before you reach for any cleaner, try soaking a rag in water, and laying it on the sticky or dirty spot. Leave it to soak for a minute or two. This is especially useful to remember when cleaning floors and countertops. No cleaners required, and equally important, no scrubbing required. Just wipe clean.

Choose an unscented soap—fewer chemicals are used to make it.

Plain soap works just as well as perfumed soap. I always buy plain glycerine soap for use in both sink and tub. (You can even use it to wash your hair.) I'm lucky enough to be able to buy it sliced off the cake in bulk at my local market. Check your natural food store or farmers' market to see if you can too. You can buy packaged glycerine soap in the drugstore or supermarket, but it tends to be very expensive—and we really don't need all that packaging. Otherwise, look for "unscented" soaps. Clean smelling doesn't mean smelling like a chemical perfume.

Try an unscented, plain cream for all parts of your body.

Some chemists say the "guckier" a cream feels, the better it is at keeping the moisture in your skin. Pure lanolin would have to rate as one of the guckiest substances. Some people put it on their hands and then wear gloves to bed. If that's not an acceptable solution for your dry skin, try a lanolin-based unscented cream.

At the drugstore, you can also buy a "base" cream that pharmacists use to make up prescriptions. It's completely white and odorless. My daughter Caroline used glaxal base when she was about three years old. I remember how she astonished her nursery school teachers by telling them that she had "atopical dermatitis." She found it soothed her dry skin without stinging at all. I've often used it for dry hands myself, since then. You don't need a prescription for the base cream, but you'll likely have to ask the pharmacist for it. It's not a brand name, and it's not on the shelves.

Any vegetable oil works as a skin cream. In a pinch— when camping, for instance—you can use a bit of unsalted butter or shortening on hands, face and lips.

Use lemon juice or vinegar for a rinse after shampooing.

This makes your hair soft and shiny. Will you end up smelling like a salad? No—you rinse it out, just like any cream rinse, and you only end up smelling clean. You can put vinegar in an empty shampoo or cream rinse container, and keep it right in the bathroom along with your shampoo. Lemon juice is best kept in the fridge.

Choose a natural deodorant.

Barry used to do a little dance and flap his arms whenever he put on deodorant, because it always stung his tender underarms. A few years ago, we discovered a wonderful product that works really well for us. It's a crystal opalescent stone which Barry and I call "the rock." We each have our own rock (you wet it and rub it on your underarm).

So far, I've only seen it marketed under the brand name "Le Crystal Naturel." Check with your drugstore to see if they carry that brand or any other. The rock is made of natural mineral salts, and is a completely natural body deodorant without any perfumes or chemicals of any kind. It's pretty expensive—$20 when I bought it in a drugstore a couple of years ago—but mine will last for several years yet. I recently saw it marketed by a natural food store for $13. But because it lasts so long, it's cheaper in the long run than buying other packaged deodorants. And Barry never dances any more when he puts on his deodorant.

Avoid spray deodorants, which use a great deal of packaging to dispense a relatively small amount of product.

Use a washcloth to remove makeup.

Paper tissues and "cotton" balls are used once and thrown away, wasting the resources to make them and filling up our garbage. Use a facecloth with warm water.

Choose the least packaged sanitary supplies.

My major complaint with sanitary supplies, aside from the use of bleached pulp, is the excessive and unnecessary use of plastics, which never break down in the environment. I'm touching on an area of strong personal preferences, but women can influence manufacturers if they think about the products they buy in terms of their environmental impact.

Choose a tampon with a cardboard applicator or no applicator. Do not flush plastic applicators. Workers who cleaned up of 157 miles of Texas shoreline found (among other plastics) 1040 plastic tampon applicators in just three hours.

I reluctantly stopped buying my favorite brand of panty

liner when the manufacturer decided to package them in a box containing sets of three liners per plastic envelope (not only did I find it environmentally offensive, I also found it less convenient). Some sanitary napkins come packaged in individual plastic bags—more plastic for our landfill sites.

An alternative to consider: sanitary napkins made without chlorine-bleached wood pulp are available in some stores. They do not have individual plastic wrappers. If you are really into alternatives, check out your natural food store or women's center for a washable, reusable menstrual sponge.

Once this year, buy your toiletries somewhere other than the supermarket or drugstore.

Your local natural food store or farmers' market may have everything you'll ever want or need for toiletries or cleaners. Look for biodegradable soaps, shampoos and laundry cleaners. In many bulk food stores, you can take your own container to fill with shampoo. Check your supermarket for new environment-friendly products (they are usually well advertised where they are available).

The Body Shop, found in major cities in North America and Britain, is well known as an environment-friendly store. They have all-natural toiletries which are not tested on animals. In Britain, they will refill your plastic containers; elsewhere, they recycle them. (The company also devotes time and money to environmental and humanitarian causes.)

Ignore the manufacturers.

One of the greatest marketers in the history of business is the person who added the last step to the instructions now found on every shampoo container: "Repeat." Unfortunately, that genius was a lousy environmentalist. Why do manufacturers recommend one cup of cleaner when a half-cup will do the job?

Experiment by starting with one-half the manufacturer's recommended quantity of soaps or detergents to find out how little you need. You may find no loss of cleaning power. We've cut back on laundry and dishwasher powder, and toiletries of all sorts with no difference in cleanliness. Our white clothes may not be whiter than white, but they're still white. Our hair, dishes, sinks and toilets are still clean. And of course, the bonus benefit in all this is that you'll save money at the same time you're being kind to the environment.

> *Let's clean up our act. Try at least one of the suggestions in this chapter this week. Invest two minutes to find out how you can make a difference painlessly.*

4

DRIVER ALERT

Is The Tank Half-Empty, Or Half-Full?

Let's admit it, we're hooked on automobiles. The internal combustion powered vehicle is little more than 100 years old, but has revolutionized the planet, and most particularly the more wealthy and developed nations. Is there anyone left in the western world who has never ridden in an automobile? Ironically, although there's one car for every two people in North America, most people in the world will never own an automobile.

Who hasn't dreamed of owning a fancier, more expensive vehicle at some time or another? Who hasn't lusted after a zippy red roadster or a prestigious luxury vehicle loaded with extras? The car as status symbol has been a part of our mythology, almost from the time of its invention. And for many people, some kind of car is a necessity. Communities are being constructed miles from our supermarkets, schools, businesses and shopping centers. Public transit is spotty at best. People who live in such communities must rely on vehicles in order to live, work, go to school or participate in the public life.

More and more people are beginning to make the connection between our vehicles and the destruction of our planet. Just as it is no longer acceptable to flaunt excessive consumption of energy in homes, to wear endangered animal furs, or to purchase tropical hardwood products, people of common sense and good will are finding it unacceptable to flaunt large energy burning vehicles. What has caused the change in attitude? We're beginning to discover the downside of our love affair with automobiles.

• **The greenhouse effect**: This gradual warming of the planet is due to the accumulation of gases in the atmosphere that trap the Sun's heat close to Earth's surface. By burning fossil fuels, such as gasoline, we release fossilized carbon into the atmosphere and create this problem. (For a longer discussion of the greenhouse effect and global warming, see Chapter 6 under "thermal energy.") United States carbon dioxide emissions are rising faster than anywhere else in the world. Canada is not far behind. Vehicles may be more efficient than in the past, but there's a greater total number of cars on the road every year.

> **The United States produces over 8 million new cars every year. Florida now has more vehicles than residents. The average North American car produces its own weight, or more—about 2 tons—in carbon every year.**

• **Acid rain**: Cars produce nitrogen oxides, which collect in clouds. When rain forms, the water is acidic. Hence the term "acid rain." California has set the standard for tailpipe emissions, although other states are now adopting stricter standards. Canadians lag well behind Americans in regulating tailpipe emissions.

Trees, lakes, fish and wildlife have borne the brunt of acid rain damage, although even stone buildings and monuments show the corrosive effects of acid rain, sleet and snow. Acid rain has killed thousands of lakes, rendering their waters unable to sustain fish, plants and micro-organisms. Acid water causes heavy metals to leach from rocks into the lakes. Fish absorb the metals into their gills, which eventually causes them to suffocate. Although these lakes may appear beautiful and crystal clear, in reality their clarity is due to the death of living organisms. Looking into an acidified lake is like looking into a test tube full of crystal clear acid.

The average car will burn three tons of gasoline in its lifetime.

Acid rain has destroyed a third of the trees of Germany's famous Black Forest, and irreparably damaged lakes and trees in Canada, Norway, Britain and other northern countries. It comes as no surprise to learn that evidence now shows that acid rain is hazardous to human beings. Children, the elderly and adults with respiratory ailments are all affected. Airborne poisons build up in the lungs, leading to lung disease and other respiratory afflictions. So we are all affected to some degree.

• **Ozone layer depletion**: Air conditioning is no longer the luxury it once was. It has become a standard feature on many vehicles, and is even offered as a "free" extra in some manufacturers' sales pitches. Depending on where they live, between 75% and 90% of North Americans who buy new cars pay extra to have air conditioning. If global warming continues at the rate it seems to be going, we will likely increase our use of air conditioning in vehicles even further. But the coolants in air conditioning, consisting of chlorofluorocarbons—known as CFCs—have a powerful and hazardous effect upon the Earth's ozone layer.

What is the ozone layer? Why should we care if it gets depleted? Ozone (O_3) is a specialized form of oxygen. High in the Earth's upper atmosphere, a layer of ozone, produced naturally from oxygen, protects us from the harmful ultraviolet rays of the sun. CFCs, which are composed of atoms of chlorine, fluorine, and carbon, break down in the upper atmosphere, releasing their chlorine. The chlorine destroys the ozone molecules in the ozone layer.

Scientists were shocked to discover that holes in the ozone layer over the Antarctic and the Arctic are growing yearly—some holes are now "continent-sized." The consequences for us on Earth include higher rates of skin cancer, increased eye cataracts, greater susceptibility to sunburn, premature skin wrinkling and aging, lower crop and timber yields, damage to marine ecology, reduced fish catches and slower growing and smaller leaved plants. Paints and plastics will also deteriorate more quickly, necessitating more frequent replacement.

The distressing news is that CFCs may take years to break down, so we do not know how much worse the ozone depletion will become, even if we stop producing CFCs immediately. While we rushed to reduce our use of CFCs in aerosol products and plastic foam products, we have not found widely acceptable substitutes for CFC's in air conditioning.

Car air conditioners use a much more destructive coolant than home air conditioners, and are subject to excessive leaking. In the past, the practice has been simply to release the old refrigerant into the air, and refill with new coolant when servicing car air conditioners. The CFCs from the old refrigerant head for the stratosphere, where they continue their relentless attack on the ozone layer. This practice may soon become illegal in many jurisdictions.

• **Smog**: Ironically, the same gas that we so desperately need in our upper atmosphere, ozone, is a major component of smog down here where we breathe. Ozone is formed in air when exhaust gases from vehicles react in the presence of sunlight. It peaks on hot summer days, especially in afternoon rush hours. If it rose to the upper atmosphere, it might replace the ozone we destroy. Unfortunately, it doesn't. We get the worst of both worlds.

About 75% of ozone-forming pollutants come from automobile exhaust.

The consequences of smog include scarring or premature aging of lungs, respiratory problems and crop damage. Anyone who has tried to breathe the "thick" air in urban centers on a hot summer day knows exactly how unpleasant smog can be.

Children's lungs can be damaged by lengthy exposure to ozone at concentrations of as little as two parts per billion, according to Robert Day, an environmental health expert with the American Lung Association. Government guidelines allow much higher levels.

• **Lead poisoning**: We are well aware of the danger of lead poisoning. Ancient Romans suffered lead poisoning from using lead drinking cups. Some have speculated that mass lead poisoning may have lead to the downfall of the Roman Empire. Explorers such as those in the famous Franklin expedition suffered lead poisoning from the lead solder in their tinned foods. House painters in the past suffered lead poisoning because of lead in paints. Even in modern homes with copper plumbing pipes, we have been told to discard the water that has been standing in the pipes overnight because lead has been used to solder the pipes together.

Initial symptoms of lead poisoning may be vague, ranging from fatigue, loss of appetite and irritability to sleep disturbances and behavior changes. More severe lead poisoning may cause vomiting, blackouts, convulsions and coma. Lead absorption over a longer period can damage brain development in young children. Adults may suffer high blood pressure. The nervous system and digestion may become impaired. Victims may become weak and suffer kidney damage.

And yet we have spewn out lead from our vehicles for years. We have used tetraethyl lead to raise the octane of vehicle fuels, without regard to the serious consequences of its use. Although we are phasing out lead in North America, it continues to be a severe problem in many Third World countries. But we are replacing lead additives in our gasoline with other additives which may prove to be harmful as well. The United States has banned MMT, an octane enhancer, but still exports it to Canada. Since almost all Canadians live within 200 miles of the U.S. border, MMT pollutes both countries. We also use aromatic hydrocarbons which may form carcinogens during engine combustion.

• **Built-in obsolescence**: The manufacture of vehicles consumes raw materials with a ravenous appetite. Besides steel, our vehicles may contain iron, copper, brass, bronze, nickel, tungsten, zinc, cadmium, chromium, aluminum, lead, plastic, fabric, foam and glass. So in addition to the huge tracts of land necessary for car manufacturing plants, we also have the landscape being ravaged for raw materials.

Changing fashions have always caused our status-conscious society to purchase beyond its necessities. Variable hemlines and tie widths have ensured a steady supply of customers in clothing stores. But vehicle manufacturers have gone a step further. By dating every single vehicle with a year, and retooling to produce a different style every year, they have cleverly ensured that your new car will depreciate considerably the minute you drive it off the lot.

And of course, it is in the manufacturer's interest to ensure that your car lasts just long enough, but not too long, so that you'll be back for another vehicle long before a decade has passed. If our refrigerators or washing machines fell apart as fast as our cars, we would raise a hue and cry. But we can't tell our 1986 refrigerators and washing machines from our 1990 ones, so our appliances can be made to last longer.

What To Do

For 10 sunny days this year, try to get to work without your car.

The first and most important thing we can do is to cut down our use of vehicles. And it may be the most difficult for some people. Our city planners have made it so much easier to take our cars downtown than to take public transit, that for some the effort seems too much. But if everyone commits to do it for just a week and a half of the next 52 weeks, we'll make a difference.

Choose a "No Car" limit.

It might be three blocks or six blocks or whatever works for your family. Our local milk store is about five blocks away. It's become the testing ground for our family. Sometimes we need milk in a hurry, and don't want to take the time to walk. Now we try to plan our needs better so that we can take a walk and get milk before it becomes too urgent. Sometimes we ride our bikes. Caroline has accepted the rule of not taking the car to the milk store, encouraged by the fact that we sometimes buy her a little treat when we get there. No treats if we drive.

Drive an extra block to the service station that will recycle your motor oil after the oil change.

Used motor oil is not only a pollutant itself, but contains dangerous heavy metals, arsenic, benzene and lead, all of which will contaminate our soils and can eventually leach into our water systems. Many garages now take their own used motor oil for recycling, and some will gladly take in any brought in by customers. Ask at your service station. In some cities, you can take your used motor oil to a central depot. Phone the local library or government office for your nearest location. If you have your oil changed at a service station, ask them what they do with their used oil. Support service stations that recycle. If you can't get your oil recycled, or if you change your own oil, dispose of it as a hazardous waste, not in your regular garbage pickup.

Use re-refined motor oil in your vehicle.

The reason we change our motor oil is that it gets dirty. It does not get used up or worn out. Re-refined oil has been cleaned and treated so it is as good as new. Most car manufacturers approve of re-refined oil for use in their vehicles. Check your warranty if you have a new car. If you have your oil changed at a service station, ask them to use re-refined oil. Not only will you preserve our environment but you'll be saving our non-renewable petroleum resources.

Try a "No Car" week.

What do we do if our car is in for repairs? We make do. See how your family can manage without the car for a week. Choose a good weather week, and reward the whole family with a special treat.

Park farther away than necessary.

If your destination is 20 blocks away, drive the first 15 or 16, and walk the rest of the way. You'll arrive full of renewed vitality after your bit of exercise, and will save the environment both coming and going. Leave yourself plenty of extra time to make your appointment.

Do at least one regular chore by bicycle.

When I first started doing my radio spots about the environment, I drove down to the studio to be there on time early in the morning. But when I did my first item about cars, I knew I would lose all credibility if I drove to the studio to tell people to leave their cars at home. So I decided right then that my bicycle would be my studio vehicle. Now, I almost always ride my bike to the studio. It's made me so conscious of car use that I've been riding my bike more and more to other places. I don't have a fancy bike, just a 15-year-old clunker, on which only two of its original 10 speeds work. But it's more than adequate transportation, and gives me some much needed exercise after spending most days at my desk.

If you always drive to the post office, you might make the post office—or the milk store, or the dry cleaners, or the bank—your bicycle destination. How about a "Bike to Work Week"? You might suggest to your employer that the company provide secure bicycle stands for employees. Any company that provides car parking should certainly provide bike parking.

. **Choose a smaller car.**

Buying a new car? Think small. A smaller vehicle costs the environment less in terms of the materials it takes to manufacture it, and it will likely be more fuel-efficient. Even luxury cars are smaller these days. If you're used to a monster car, you might discover that a small car is a treat. After years of big luxury cars, Dad complained mightily that the new cars had a small wheelbase. When he did get a smaller car, he became an instant convert. He loves the convenience of zipping around and the ease of parking. The reduced gas bills are a bonus.

Choose a fuel-efficient vehicle.

If you're getting less than 25 miles per gallon, it may be you're driving a big car, you're going too fast, or your vehicle needs a tune up. Aim for at least 35 miles per gallon (eight liters per 100 kilometers) if you're buying a new car. Of course there are other factors to consider when you're buying. But fuel efficiency should be near the top of the list. Your library should have a copy of the government ratings. Or try the telephone book for the government transportation office nearest you.

For fuel economy, allocate vehicles according to distance driven and fuel consumption of each vehicle.

Two car family? If you have his and hers cars, the person who drives farthest gets the most fuel-efficient car. No arguments. It makes no sense for the one who drives

10 blocks to take the mini-car while the one who drives five miles takes the gas-guzzler.

If you've been doing it the other way around your family will notice an immediate improvement in gas bills, just by switching vehicles.

Choose radial tires for better gas mileage.

Improved tire performance will be kinder to the environment in a couple of ways. First, you won't have to replace your tires so often, which means reduced waste of natural resources from manufacturing through to ultimate disposal. Second, you'll get better gas mileage, which will mean reduced waste of carbon fuels, and all which that implies. Because of their construction, radial tires grip the road more efficiently than bias tires.

Take two minutes to check your tire pressure twice a month, especially in changing weather.

Inflate your tires to the manufacturer's recommendation for best fuel mileage. Do not underinflate, which causes a greater drag on the surface and reduced efficiency.

Have your tires properly balanced to correct uneven weight distribution.

If your tires are out of balance, your vehicle will vibrate, and your gas mileage will go down.

Check your vehicle wheel alignment.

A misalignment of just 1/2 degree in your wheels is equivalent to dragging a tire sideways 8.7 miles for every thousand miles you drive. Besides giving poor fuel consumption, incorrect alignment will greatly decrease the life of your tires.

Let your fingers do the driving for two minutes, to see if there's a tire recycler in your neighborhood.

Tire recyclers may pay for tires delivered to them, or will pick up used tires. It is becoming increasingly difficult to recycle tires because of the materials used and the methods of construction of high quality new tires, but you may find a recycler near you.

Add two minutes to your trip: slow down to save gas.

Not only will you save on fuel consumption, but your vehicle will produce a reduced quantity of poisons such as carbon monoxide and hydrocarbons. Consider: If your journey takes you an hour at 70 miles per hour, it will only take an hour and 10 minutes at 60 miles per hour. On shorter distances, the time difference will be negligible. Fuel consumption increases by 20% when driving 70 miles per hour rather than 55 miles per hour. Speed not only kills—it is hard on our environment.

Start the car two minutes later than usual. Wait until everyone is ready to go.

Do all your pre-flight clearances before you start the engines. That includes adjusting your groceries, your parcels, your coat, your gloves, your sunglasses, your rear view mirrors, your seatbelt, your cassette tapes and your passengers.

Every extra minute we spend with the motors idling means more poisons in the air and more carbon fuels burning. Experts agree that even in cold weather, a few seconds is all that's needed to warm up the engine before pulling away—slowly. Your heater may not blow warm air immediately, but your engine is ready to work. And please don't leave your car running while you wait for your passenger to run into a store. Turn it off.

Reduce your use of air conditioning in the car.

Air conditioners use extra fuel. Do we really need to be cooled to the point of wearing sweaters in the summer time? Play a game with your family or yourself to try using it less. Get the kids involved. See how far you can drive without the air conditioner. How about to the next red building? Can you make it until someone spots a licence plate with a "Z" on it? Give a one minute blast of cold air for the first person who sees a female bicyclist with a black helmet.

If you're driving for 20 minutes or less, you may be able to do the whole trip without air conditioning. All participants receive an ice cream cone reward. Complainers get zip.

Post up a car pool notice on the office bulletin board.

If you live in a big apartment building or work in a big office tower, there's a good chance you could car pool with someone. State the major crossroads nearest your residence and work place, your departure times, and your office telephone number. If even two people share a vehicle, both will save half the expense of gas and parking. Consider car pooling to church, to school, to meetings, and to extra-curricular classes and activities.

Some parking lots give a reduced rate for vehicles containing more than one person. If you work for a big company, ask the boss if the company will subsidize some "perks" or bonuses for car poolers—good public relations for the company.

In two minutes, you can use the phone book to find a service station that will let you clean and reuse your antifreeze.

Many municipalities have laws governing the dumping of antifreeze, but according to industry insiders, the laws are ignored as often as not. One service station owner told me that antifreeze regularly goes down the drain, in spite of regulations requiring that it be put into a sump and disposed of as hazardous waste. Used antifreeze itself, consisting of ethylene glycol mixed with water, constitutes a hazardous waste. But worse, it also contains metals and other sludge waste which will contaminate our water supply when it's dumped down the drain. It is these contaminants that necessitate the regular change of antifreeze in our vehicles.

A new process, available at more and more service stations and at many Ford dealerships, lets a mechanic flush your old antifreeze into a cleaning machine, remove the corrosives and metals, and return the antifreeze to your radiator, as good as, or better than, new.

Better? Yes, because before this cleaning process was developed, about 25% of the antifreeze stayed in the car when you thought it all had been drained. With the new method, pressurized air forces 95% of the antifreeze out of the system for cleaning. New additives are put back in to restore lubricating and anti-corrosive qualities. So ask your service station if they have the system yet, and if not, when will they have it?

Antifreeze producers, not surprisingly, have bad-mouthed the system, predicting car trouble, but Ford believes in it and hopes to have it installed in most of their dealerships. Check your new car warranty. One service station owner told me they had cut their antifreeze order from 1320 to 317 gallons per year after installing the recycling machine. No wonder antifreeze producers don't like the idea.

Clean and reuse your vehicle air conditioner refrigerant.

Recycle your air conditioning refrigerant. Car manufacturers have not rushed to replace the damaging R-12 (refrigerant 12) which is used in automobile air conditioners. It's considered one of the worst offenders for damaging our ozone layer. Our car air conditioners leak as often as not, and may require three or more refills of coolant during their lifetime.

Many service stations now have the equipment to capture and recycle the coolant. Ask your service station to capture your coolant when you have your air conditioning serviced. If they don't have the equipment, look for a place that does. General Motors is working toward putting the equipment in hundreds of their dealerships. While you're at it, have your air conditioning system checked every year, and purged of water vapor. Do not top up the system with canned refrigerant (now banned in many jurisdictions). Repair any leaks, rather than just replacing lost gas.

> **Drive to the station that sells an alternate fuel, such as ethanol-blended fuel or "gasahol."**

If alternate or "renewable" fuels are available where you live, try them. (Renewable fuel is made from a resource which will not become depleted if properly managed.) Here are a few to consider.

• Ethanol: One such fuel is ethanol, an alcohol fuel made from a variety of biological materials, such as low quality corn and grain, even cornstalks or wheat straw. The beauty of ethanol is that the carbon it gives off as it burns is no more than the amount of carbon the grain absorbs as it grows, thus making the process a "closed loop." Theoretically, the burning of ethanol does not add to the greenhouse effect, and may even reduce atmospheric carbon dioxide levels in certain circumstances. In practice, ethanol is usually blended with gasoline.

The use of alcohol fuels also reduces exhaust emissions of unburned hydrocarbons and carbon monoxide—environmental villains in the formation of ozone.

Ethanol-blended fuel also enhances the octane or performance level of vehicles. It may be used in any vehicle without modifying the vehicle, and has been approved by every vehicle manufacturer. In fact, some new vehicle manuals, particularly those for General Motors 1990 models, are recommending its use.

Local ethanol manufacture would decrease dependence on oil supplies, ensure a steady market for agricultural grains, and increase employment in rural communities. A valuable by-product of ethanol production is a high-protein animal feed, making the process an even greater boon to agriculture.

Ethanol fuels are widely available in the United States, because renewable fuel production and distribution are subsidized by the government. Unfortunately, gasohol—as blended ethanol fuel is often called—is not widely available in Canada. Only the four western provinces, British Columbia, Alberta, Saskatchewan and Manitoba are served by Mohawk Fuels, which is the major manufacturer of ethanol fuels. Look for their service stations in western Canada.

In 1988, ethanol blend sales were 8% of the total gasoline sales in the United States. Americans have driven over 700 billion miles on ethanol blended fuels since 1979. Canadians have driven about 3 billion miles.

• Propane and natural gas: These are often widely touted as being good alternatives to gasoline. In fact, they do burn much more cleanly, contributing far less carbon monoxide, hydrocarbons and particulates than gasoline. But natural gas does give off methane, which contributes to the greenhouse effect. Although they are in good supply at the moment, these, too, are carbon fuels, and not a renewable resource. They were formed millions of years ago. We cannot manufacture them. Like gasoline, and the petroleum from which it's made, when they're gone, they're gone.

Another drawback to the use of these fuels is that vehicles must be retrofitted to burn them, adding $2000 to $2500 to the cost of the vehicle. And fuel stations may not be available in all areas.

• Methanol: Like ethanol, methanol is an alcohol fuel. It, too, burns cleanly, and enhances octane levels when blended with gasoline. To burn methanol "neat," without blending, vehicles must be modified. This method produces fewer oxides of nitrogen, and thus contributes less to acid rain than does the burning of gasoline. Carbon monoxide emissions will also be reduced, although other emissions may add to ozone formation. There is no effect on carbon dioxide emissions as compared to gasoline.

Methanol-blended fuels may not perform as well at high

speeds as ethanol blended fuels. Furthermore, a co-solvent such as ethanol must be added to them. Methanol is derived from natural gas, a non-renewable fuel.

• Diesel: Diesel fuel is a petroleum fuel. It, too, is non-renewable. However, the diesel engine is more efficient than a regular gasoline-burning engine, particularly for city driving, and conserves fuel. The diesel engine will probably last longer, which implies reduced use of our resources in manufacturing.

However, diesels are notorious for the soot and fumes they produce which may be harmful to our health. Diesel trucks are worse than cars in this aspect, although both are undergoing changes which may improve the situation. If the soot and fumes are brought under control, diesel cars will be a vehicle of choice for environmentalists.

> *Driving a car is a privilege and a luxury for two or more people to share. Every five years, we add about 100,000,000 more vehicles to our planet.*
>
> *Explore your world on foot or by bicycle. Walk, bicycle, or take public transit to work once a week, once a month, or at least once in your life.*

Chapter 5

EXCESS PACKAGING

Boxes In Boxes In Boxes?

It's almost impossible to buy certain things without packaging these days. We can't buy razor blades, batteries, soap, screws, or toothpaste, without having to throw away a package as soon as we get home. Have you ever bought pantyhose or nylon stockings without buying a throwaway package (and a cardboard insert) at the same time? What about larger items like electrical appliances (everything from toasters to refrigerators)? Ever try buying one without a box?

Some things must be packaged. We don't carry home chocolate chips in our hat, or flour in our skirts—not that we couldn't, but it's most unlikely that we will. (There are alternatives to accepting the pre-packaged chocolate chips and flour. We'll discuss that below.) Packaging for tiny, powdery or liquid items is packaging for convenience of carrying.

Some things are packaged for other reasons, such as

safety (medicine and razor blades), preservation (juices and canned goods), and protection of fragile items (light bulbs). I suppose we should be grateful to the packaging industry for providing us with the containers that serve these purposes. But somewhere along the way, packaging became so wildly successful that it started spilling into areas where it served a less than lofty purpose.

Since 1960, the amount of packaging in garbage has increased by 80%.

The packaging industry sold wholesalers and retailers on the idea that more packaging would cut down on theft, handling, breakage, waste, spoilage and tampering. At the same time, they increased the number of options in sizes, shapes and varieties of packaging. Packaging became a selling tool—package designs became part of the advertising of any given product. "New and improved" often means little more than a new and improved package. And we have started seeing the package within a package on more and more items.

Now for convenience, you can buy cheese slices, each individually wrapped in plastic, and all those tiny individual plastic packages wrapped within a larger plastic wrap. For convenience, you can buy bars of soap, each individually wrapped in plastic or paper, often with paper liners inside that wrapping, and sets of four or six bars wrapped together in a larger plastic wrap. You can buy cassette tapes, encased in a hard plastic container, which in turn are encased in cardboard wrappers, which are often covered with a plastic overwrap. Over-the-counter cold and headache remedies have about as much packaging as is humanly possible—sometimes you get a paper backing over a foil backing with individual bubble capsules for each pill, and the whole works enclosed in a cardboard container.

What happens to all that packaging? I'm pretty good at thinking of alternate uses for just about any kind of junk, but some of that packaging defeats me. I can't think of a secondary use for cheese slice wrappers or the bubble packaging from batteries or the cardboard backing for garden hose fittings. All that stuff becomes garbage the instant it leaves the store.

In North America, packaging from consumer products makes up as much as 50% of the volume of municipal solid waste. Canadians, followed closely by Americans, make more garbage per person than anyone in the world—four pounds per person every day. In a single year, Americans generate over 266,000,000,000 pounds of garbage.

How much of your garbage comes from disposable packaging?

The environmental costs associated with packaging may not be evident at first glance. The packaging industry is a manufacturing industry. In order to have glass, plastic, paper, or metal packages, we must first extract the raw materials from the earth, depleting our natural resources. Then we must generate energy to manufacture and shape those materials, sometimes creating toxic by-products which go into our water or waste stream.

Once we have manufactured the packages we must transport them, first to the wholesaler who wants to fill the packages, then to the retailer who will sell the filled packages, next to the consumer who buys the products and packages together, and last to the landfill or incineration site, where the empty packages are finally out of sight, out of mind.

But the costs don't end there.

Some packages may be inherently toxic. Dioxins formed in the manufacture of bleached paper products have led to concerns about the possible contamination of milk and other products sold in cardboard cartons.

Some packaging (and other garbage) may create toxic by-products as a result of either degradation in landfill sites or incineration. Landfills may leach toxic waste into ground water or streams. Incineration can generate both toxic air emissions, and toxic ash waste.

> **Plastic debris is hazardous to wildlife. According to the World Society for the Protection of Animals, each year more than two million birds, turtles, whales, dolphins and seals become entangled in floating plastic debris, or mistake it for food.**

Litter, of which up to 65% may be packaging, destroys the natural beauty of our parks, beaches, streets and wilderness.

Our landfill sites, which are rapidly being filled to capacity, use land that could otherwise support wildlife or agriculture. Once full, the use of the land is restricted by the potential problems arising from the assorted garbage.

All of the costs of dealing with these problems will be borne by either the consumer or the taxpayer or both.

What To Do

Take less garbage home from the super-market.

Let's start at the source—your empty shopping basket. Instead of recycling more, let's create less garbage in the first place. How much of any given product is packaging? How much of it is necessary? How much do you really need an over-packaged product? Is there an alternate product with less packaging?

Say I'm buying a hacksaw blade or an electrical outlet. At the hardware store there are cardboard-backed, plastic bubble packages of each available. There may also be some bins with loose hacksaw blades or electrical outlets. I'll always choose the loose ones. They're often the same price, but I've even been known to pay more for goods without the packaging. Call me a fool, but I'm thinking of the hidden costs of packaging. So I assume I'm getting more for my money when I buy without the package.

In the same vein, look at grocery purchases like cheese singles. Just for a change, try the deli counter of the supermarket, and ask them to slice some cheese for you. You can still have cheese slices, but without all that extra packaging. As a bonus, you might even find you prefer the taste of the deli cheese.

Look at sports socks. Do you have to buy a plastic package of six pairs? Can you buy the same six pairs wrapped in a band of paper? Even better, can you buy the same six pairs loose without any wrapping at all?

Look at cosmetics. Can you find an equally acceptable

handcream that comes in a single container, instead of one that comes in a jar shell within a jar within a box?

Take your own shopping bags or boxes to the supermarket.

What a golden opportunity to cut down on waste. Suppose you want to buy five bananas, four apples, a bunch of grapes, and a loaf of bread. How many plastic bags will it take to get out of the store? One for the bananas, one for the apples, one for the grapes, one for the bread, and one or even two big ones to carry the whole works in.

Of course that would be a very light day's shopping for most families. Multiply those bags by the amount that you actually buy in any given shopping trip, and again by the number of shopping trips you make in a year. You'll see that you can end up with literally hundreds of plastic bags in the space of a year. (In fact, I'll bet there's a stash of plastic bags somewhere in your house—the bag drawer, the bag cupboard or the bag bag. And there are probably so many bags in it that the drawer won't close.) And you're just one consumer.

Instead, try this scenario: Go to the grocery store armed with your own supply of plastic bags, both the transparent produce bags and the regular supermarket bags with the carrying handles.

You may be surprised at how few bags you require. For example, I seldom bag my apples or other large fruits or vegetables separately from my other groceries. I just put them straight in the cart, along with my oranges and cauliflower. Sometimes if I'm buying a lot of apples, I will use one of the bags that I've brought with me, just for convenience. Usually I don't bother for a half-dozen apples.

Don't worry about seeming peculiar or unusual. When I first started taking my own bags to the supermarket, about 15 or 20 years ago, I sometimes had to argue loud and long just to get them to allow me to use my own bags. I can remember almost being tossed out on my ear by one supermarket manager who thought I was trying some kind of thievery scam.

Nowadays, taking your own bags is commonplace. In fact, many stores encourage it, because you're saving them the cost of the bags. Owners of small stores always thank me for bringing my own bags and boxes.

Some supermarkets have made the connection that "free bags" doesn't really mean free. Somebody has to pay for them. Some stores have started charging customers for bags. Alternatively, some pay customers 3¢ or 5¢ when they bring their own bags to the store. Some stores provide cardboard boxes for customers. Others provide plastic carry home boxes for a nominal returnable deposit. Any of these choices will make a difference in the amount of waste plastic and paper that's smothering the earth. I've been taking the same four or five boxes to the grocery store for at least a couple of years, and they haven't worn out yet.

Carry things home from the store without a bag.

What if you forget to take your bags with you when you go shopping? Or if you don't have a bag in the car? If it's just one or two items, like milk and bread at the corner store, I simply tell the clerk, "No thanks, I don't need a bag." I carry the things out in my hands. I've done it for years without problems.

Some people think that if the item is in a bag it's proof that you've paid. I never worry that someone might think I'm stealing because I don't try to hide the merchandise. I just walk out with it right in front of me. Usually I carry my receipt dangling between my fingers. I've seen people buy single items like razor blades or toothpaste, all of which are over-packaged already, and meekly stand there while the clerk puts it in yet another package, which they carry home and throw in the trash. They could just as easily have popped it in their pocket or purse along with the receipt, or carried it out of the store without a bag. No wonder we have so much garbage.

> **Go back to a permanent razor. The two minutes you spend changing the blade will help.**

Throw-away razors create mountains of plastic garbage. The blades for permanent razors require less of our resources to manufacture, and create much less garbage.

> **Take a moment to mail a wasteful package back to the manufacturer, with a letter telling them we don't need it and don't want it.**

Your letter should tell the manufacturer that you like their product, and don't want to stop buying it, but you'll do so if they continue to over-package. Then be prepared to follow through. Money talks, and manufacturers will listen.

Praise the manager of a business that doesn't package excessively.

Sometimes we take for granted that something we like will always be available. But if business owners don't know that one of the reasons you buy from them is their restraint in packaging, they might be tempted to follow the trend to more packaging.

The same idea holds for bigger supermarkets, although it may be harder to get the message across. It never hurts to say to the produce manager, "I'm glad there's no packaging on the broccoli this week," as you put it in your cart.

Reject unnecessary wrapping on fruits and vegetables.

Some things really drive me crazy, but the thing that bugs me the most is finding coconuts wrapped in plastic. Coconuts have shells that can survive long ocean voyages, need a hammer to crack, and protect the edible part better than any wrapping devised by humans. Yet I see them in the grocery store wrapped in plastic. Next on my list of grocery store absurdities is the bananas on a foam tray covered with plastic wrap. Bananas come in a beautiful biodegradable (and compostable!) zipper package, straight from the trees. They certainly don't need plastic on them.

Try your local markets or smaller stores. You may be able to find lettuce that doesn't come wrapped in plastic, or broccoli that doesn't have a foam tray and plastic wrapping.

Make choices in this order: Reduce, reuse, recycle.

In the soft drink aisle, you may be confronted with a bewildering variety of choices in packaging: returnable glass bottles, non-returnable glass bottles, recyclable plastic bottles, non-recyclable plastic bottles, and recyclable cans. How to choose?

The first choice would be to reduce our consumption—that is, not to buy the soft drink at all. Having rejected that choice, since we came to buy soft drinks, we move on to the second choice: reuse. Right away, we know that the only package that fits that choice is the returnable glass bottle. If a glass soft-drink bottle is returned and reused just a few times, it outstrips all the competition in terms of its environmental friendliness. If your store doesn't stock returnable bottles, choose the container which you can recycle. In some communities, cans, glass and plastic bottles are all recycled. In that case, there's not a lot to choose. Metal, glass and plastic all cause environmental problems in their manufacture. Glass would be my first choice for its complete recyclability, but plastic is lighter weight for transporting, thus causing less wasted fuel energy. Recycling steel uses 90% less natural resources than manufacturing steel with iron ore and coal. Of course, if your community only recycles one of the choices, by all means, choose that package.

Buy bulk foods whenever possible.

Another great place to cut down on waste packaging. Remember to take your own bags to be reused over and over again for everything from flour (saves carrying it in your skirt) to cat food to spices to nuts and chocolate chips (saves carrying them in your hat). We sometimes buy our milk in plastic bags. Rinsed and dried, those sturdy milk bags are ideal for carrying bulk foods. I also take my own clean hard plastic containers for bulk foods like peanut butter, and honey, and deli foods like olives. Yogurt, margarine and cottage cheese containers all make excellent reusable packages for such "wet" groceries.

Keep a shopping bag in the car for unplanned purchases.

Sometimes you get a phone call just before you leave work, "Can you pick up some milk on the way home?" Or you suddenly remember that you used the last slice of bread for your lunch sandwich, and have to stop at the bakery. I usually stash a bag or two under the front seat, where they're never seen, but always handy for emergencies.

Buy the largest possible packages of nonperishables.

Sometimes I see people with very small packages of rice or powdered milk or some other non-perishable in their shopping cart, and I wonder why they bother. If you can afford it, and if you have the storage room, choose larger packages.

Buy bulk hardware items whenever possible.

It's hard to find things not individually packaged any more. When you do find a hardware store that will sell you the nine loose screws that you need, instead of 12 screws and a throw-away package, stick with them like gold. We have to encourage stores to be environmentally friendly, and that means spending our money there. And again, you'll certainly save money by buying nails and washers and other hardware fittings in bulk, rather than in packages.

Choose a tube rather than a pump toothpaste.

The hard plastic pump represents little more than fancy merchandising—all you want is the toothpaste itself. It's pretty well impossible to find toothpaste that's not over-packaged, but the pump type is not biodegradable, and takes up more of our resources and landfill space than the other over-packaged types.

Buy a lipstick brush.

If you don't throw away the ⅓ of your lipstick that's in the bottom of the tube, you'll cut back on the manufacturer's use of chemicals and waste of packaging.

Use a battery charger and rechargeable batteries.

Batteries, because of the heavy metals, acids and other chemicals in them, are one of the largest contributors to toxic waste. They should never be thrown in with your household garbage. They should be saved for a toxic waste collection, or taken to a toxic waste collection depot.

That seems like an awful nuisance when you've just got a couple of little batteries, and it would be so easy just to chuck them out. Who would ever know? But if you're a heavy user of batteries, you could be doing serious damage to the environment. Besides, every time you buy batteries, you have to get one of those non-recyclable cardboard and plastic bubble packs to throw away.

There is a better way. A battery recharger costs around $15 to $20. Rechargeable batteries, although more expensive initially than non-rechargeables, will last for years. The initial investment is well worth the price. When my daughter Caroline was three years old, we bought her a battery-operated cassette player. We also bought four rechargeable batteries. Five years later, my daughter is still using those same four batteries. We've recharged them dozens of times, and they still work perfectly. Without the charger, we would have bought dozens of sets of batteries, plus all the packaging that comes with them, and thrown them all out. Of course we recharge our other household batteries too. Both the charger and the batteries have paid for themselves many times over.

Use a garbage can instead of plastic garbage bags.

Garbage bags were invented to fulfill a need that didn't exist. I have to laugh when I see people spending money on something that is made specifically to throw away. Well, I would laugh if it weren't so lamentable. Again, the initial investment will pay for itself many times over.

Garbage cans last for years and years, never clog up our landfill sites, and keep out marauding animals besides. And now that you're cutting down on all that packaging, you won't need a very big garbage can. If you live in an apartment building that requires garbage to be bagged, try to use bags that you can't avoid buying, such as rice or flour sacks, milk bags, or large pet food bags. Use plastic bags that come from the manufacturer with large purchases such as pillows or lampshades. Avoid buying plastic bags just for garbage disposal.

> *Manufacturers like to hear from their real customers, not just the survey groups. Use your purchasing power to protect the environment. Tell manufacturers and store owners you're avoiding products that waste our resources and increase the strain on garbage services. They'll get the message.*

Chapter 6

SAVING ENERGY ALL YEAR 'ROUND

Why Save Energy?

In its simplest terms energy is the capacity for doing work. And that's as good a way as any to understand why we should save it. Of course the energy we're talking about here is electrical and fuel energy, the "workers" of our modern civilization.

When we turn on a light, a clothes dryer, a furnace, or a computer, we are "burning" energy. We have taken some fuel (coal or natural gas), or a process (nuclear or hydroelectric), and used its energy to do work for us. When fuel has been burned, it no longer has the capacity to do work—it is no longer energy. To get more energy, we have to burn more fuel.

Electricity is a "secondary" source of energy. It must be manufactured from primary sources. Our electricity comes from three main sources: nuclear energy, thermal energy (energy from burning coal or other fossil fuels), and hydraulic (energy from water dams). Each of the major sources of energy poses its own environmental problems.

• **Nuclear energy**: While undoubtedly a cleaner source of energy than the other major sources, nuclear power is troublesome for several reasons. One of the big drawbacks is that there is no effective long-term solution to the problem of storing the lethally dangerous nuclear wastes. Spent nuclear fuel remains highly radioactive for hundreds of thousands of years. We have had less than 30 years experience in storing nuclear waste.

The latest proposal suggests burying it in containers made of titanium and copper. It is expected to keep the used radioactive fuel from contact with groundwater for at least 500 years. (Of course it hasn't been tested for 500 years—it's only just been invented.) If it doesn't live up to its expectations, well, too bad, we tried. And if it does work for 500 years, what then? Are we supposed to be happy that our descendents of five centuries hence will have to look after the problem of a fuel that will still be radioactive for hundreds of thousands of years? After all, we won't be around to worry about it.

Suppose our ancestors had discovered nuclear power in the middle ages, and had decided to bury their radioactive fuel—possibly on the site of a city not then in existence—say Calgary, Montreal, or Boston. Today we would have to contend with still-hazardous radiation from the used fuel of 500 years ago. Would Calgarians today be finding their water unsafe to drink? Would Montrealers have higher-than-average cancer rates? Would Bostonians be subject to birth defects? We simply do not know.

It's immoral to saddle our descendents with the risk. Yet if we leave the contaminated materials above ground, we face the risk of accidental exposure ourselves.

Another problem with nuclear generated power is the risk involved in uranium mining—essential to the manufacture of nuclear energy. Uranium mines discharge highly toxic radium 226, lethal radon gas and radioactive lead. The radioactive "tailings" or leftovers from the milling process must be contained with dams, hills or levee walls, or placed into the mined-out open pit. Water which settles out of the tailings is discharged back into the environment with barium chloride and other chemicals added to remove the radium.

Canada is the world's leading producer of uranium. Over 130 million tons of highly toxic uranium tailings from mines are now sitting exposed to the environment in many locations across Canada.

Uranium mine workers are exposed to more radiation than any other workers in the nuclear industry. Radioactive pollutants are found in plants, animals and people living around uranium mines. We do not yet know the effects on human beings who consume fish or game from around uranium mines.

Finally, although we are assured that new North American reactors are very different from the reactors that caused the disasters at Chernobyl and Three Mile Island, we also know that even the best-intentioned designers and builders of nuclear reactors are only human, and subject to the same errors and foibles as the rest of us. Nothing in our world is totally safe and foolproof, but the consequences of nuclear accidents are horrendous in comparison to other kinds of mistakes we make.

The accident at Chernobyl released 90 times the radiation of the bomb dropped on Hiroshima. The worldwide consequences may never be known.

• **Thermal energy**: Our major sources of thermal energy are coal, oil and natural gas. We can burn these fuels directly, that is in our vehicles or in our furnaces to heat our houses, or we can use them indirectly, by burning them to create electricity, which we then use to heat our houses and run our appliances.

These are all "fossil" fuels. They were created from living materials thousands of years ago, and fossilized into their present forms. When we burn them, we release the carbon dioxide (CO_2) that has been trapped in them since their formation. This leads to the greenhouse effect we hear about so often. Carbon dioxide and other gases in our atmosphere act in the same way as the glass ceiling of a greenhouse. The sun's heat can reach the earth through the "ceiling" of gas, the same as it enters a greenhouse. But the heat radiating back from the surface of the earth is unable to escape through the same ceiling, leading to a gradual increase in the earth's temperature.

The problem is exacerbated by the disappearance of the tropical rainforests. Not only does the burning of the forests release CO_2, but the trees themselves absorb large quantities of CO_2 as they grow. When the trees are gone, their benefits die with them. So we have more CO_2 being produced, and fewer trees to consume it.

By the year 2030, the earth will be warmer than at any time in the past 120,000 years.

The effects of this global warming will be earth-shaking. Fertile regions will become arid desert. Water supplies will decline significantly in rivers, lakes and reservoirs. Pollution in lakes will become more concentrated. Crop production will fall drastically.

Forests will disappear, and reforestation projects will fail due to changing soil quality. Wildlife may be unable to adapt to changing conditions. Lakes and rivers may become too warm to support the species that live there. Wetlands will disappear, leaving numerous species of birds and other wildlife homeless. Fertile coastal areas will be flooded. Some cities will become uninhabitable.

How does all this relate to our energy consumption? It's simple. The more energy we consume, the more we burn fossil fuels of all types.

Coal is a particularly bad culprit in the greenhouse effect. But its impact on the environment doesn't end there. Burning coal also produces sulphur dioxide (SO_2), a direct cause of acid rain. (Scrubbers can help reduce sulphur emissions, but the cost has delayed installation on a mass basis.) Acid rain is responsible for serious damage to numerous lakes, fish, forests, crops, and even buildings and monuments. So the more coal energy we use, the worse our acid rain problem.

We release over 100 million tons of sulphur dioxide worldwide each year.

• **Hydraulic energy:** While some of our hydroelectric power comes from natural waterfalls such as Niagara Falls, or from fast-flowing rivers, much of it is generated from dams we have constructed for that purpose.

The flooding that occurs behind a dam has tremendous environmental consequences in any region. Fertile land is lost forever. Flooding destroys forests and wetlands, along with their wide diversity of plant and animal life. Whole species may be wiped out in a given region. Farming, hunting, fishing and settlement are seriously disrupted or destroyed.

In the James Bay water diversion, the huge reservoirs are already contaminated with mercury released from flooded land. Environmentalists are protesting the next phase of the project, hoping to limit further damage.

The National Audubon Society has warned that environmental damage from the James Bay hydroelectric dams in Quebec would threaten several species of birds with extinction, and pose a risk to beluga whales, polar bears, seals and fish.

The Balbina Dam in Brazil has already drowned 1550 square miles of virgin rainforest, killed thousands of animals such as monkeys, jaguars, turtles and fish, poisoned at least one river, wiped out hardwood and fruit tree species, eliminated the source of livelihood for thousands of people, uprooted indigenous tribes, and provided a breeding ground for malaria-carrying mosquitoes. The Narmada Dam in India is the largest ever planned. It calls for 573 communities to be flooded and 1,000,000 people displaced, while forests and farmlands will be drowned under 50 feet of water.

New dams change forever the ecology of a region. Once a region is flooded, we cannot recapture the lost lands or species. And there is no way to alleviate the environmental damage. Survival depends on a complex and interactive web of organisms. We cannot dig up or capture threatened plants or animals and transplant them to another area. They will not survive. Proposed dams in Saskatchewan and Alberta have been fiercely opposed by environmentalists for just these reasons.

What good will conserving energy do if we've already got nuclear plants, coal-burning plants, hydroelectric plants and oil refineries?

Since we can't store electrical energy, we have to build facilities to meet our maximum needs at any given time. If we all turn on our air conditioners, clothes dryers, vacuum cleaners and stoves at the same time, and leave all our house lights on while we're at it, we'll need more electrical plants to meet that peak demand.

On the other hand, if we conserve energy, we may be able to make do with the plants that we already have. Wouldn't it be great if we could actually shut down some of those plants because we don't need the energy?

If we all use our oil burning furnaces inefficiently, we'll need more and more oil refineries. And the more we burn, the worse the greenhouse effect. The more we save energy, the less acid rain, the less global warming, the greener our planet.

Obviously, in a perfect world, we would all switch to

passive or environmentally benign sources of energy, such as solar power. It's unrealistic to think that this will happen very quickly. But we should not continue destroying our planet with wanton use of its resources.

The World Energy Congress has forecast that in just 30 years, the world will be using up to 75% more energy than it did in 1985. But it also made a startling observation that offers hope: energy efficiency is the largest single energy source available for future needs.

If we save energy, we are, in effect, making new energy available for necessary uses. If everyone becomes energy efficient, we may not need new power plants for the increasing demands of a growing world population.

What To Do All Year

Turn off lights when leaving the room.

Lighting accounts for a large portion of our energy bills. Let's settle the question once and for all: does it take more energy to turn lights off or leave them on? Incandescent bulbs take slightly more energy to turn on than to leave burning. Here's a rough guideline to help you decide whether to turn the lights off when you leave a room: If the bulb is 100 watts or more (or two 60-watt bulbs or more), no matter how long you'll be out of the room, turn off the lights. If the bulb is lower wattage, or a fluorescent light, and you will be out of the room for no longer than three minutes, you may leave it on. As a general rule, since we often stay out of the room longer than we expect to, we should turn off the lights when we leave.

Try new compact fluorescent bulbs instead of incandescent lightbulbs.

New developments in fluorescent lighting make energy-efficient bulbs a wise choice. Compact fluorescents, which consume about one-fifth the energy of ordinary light-bulbs, now come in the shape of incandescent bulbs, and can be screwed into a regular lightbulb socket.

A 9-watt fluorescent can replace a 60-watt incandescent bulb; a 13-watt fluorescent gives the same light as a 75-watt incandescent. Although the new bulbs are a great deal more expensive than ordinary lightbulbs (ranging from $18 to $30 each), they will last from 10 to 20 times longer. Combined with the energy saving, they will certainly pay for themselves.

Fluorescent bulbs come in a wide range of colors today, imitating warm incandescent light or daylight, so you won't be stuck with that washed out white light that reminds you of a service station restroom. They're slightly larger and longer than incandescent bulbs, which makes them too big for many table lamps, but I've used them successfully in hallways, porches, basements and ceiling fixtures with 10″ or larger globes. If you leave any light on overnight for security, it should be a fluorescent.

Try your local hardware store. If you don't see the compact fluorescent bulbs, ask for them. Perhaps you can get together with some neighbors or with your co-op to order a large enough quantity to make it worthwhile for your store to stock them. Major lighting and electrical distributors all have the bulbs available, and usually at a lower price than hardware stores.

Turn down the temperature on your water heater.

A setting of about 50° C or 122° F is adequate. If you don't have temperature markings, try setting your water heater somewhere between warm and hot. If you're going on holiday for a week or more, set it at low, or "vacation" (pilot light only) setting.

Experiment with cold water washing in the laundry.

Do you wash and rinse your clothes in hot or warm water? It takes a lot of energy to heat that water. I've done a bit of experimenting with hot, warm and cold washing, and have come to these decisions: I still use warm water to wash my whites and light colored clothes, but always use cold to wash dark clothes. And I always use cold water to rinse, whether whites or darks.

I've discovered that my ordinary laundry soap, as well as regular detergent, works equally well in any temperature of wash water. You don't need special detergents. Sometimes I dissolve the soap in a little warm water at the beginning of a cold water wash, but even that's not really necessary. I really can't tell the difference between clothes washed in hot, warm or cold water, so why waste the hot? And of course, I try to run the washing machine with full loads only—to save both energy and water.

Beware of over-drying clothes.

Using the clothes dryer less will cut back significantly on energy consumption. If you throw your damp sweat-shirt and jeans over the back of a chair, they'll be dry by morning. It's wasteful to run the dryer an extra 20 minutes just to finish off those two bulky items. Sometimes you can just hang up clothes on a hanger if they're still damp, and they'll dry in an hour or so.

I've seen people run their dryers for 80 or 90 minutes per load, just to get those last few items bone dry. The average dryer load at our house takes 35 to 40 minutes. At that time, I still have dampish towels, and possibly a few damp sports socks or jeans. But our clothes never have "static cling" from overdrying. No fabric condition-ers are required. Anything that isn't completely dry gets thrown over the wooden clothes rack, or hung up to dry. A folding wooden clothes rack or a clothes line is an extremely practical and energy-saving device.

Promote environmental awareness at your club.

Do you work at or belong to a health or fitness club? Paying a fee doesn't entitle you to abuse the environment. Does your club offer disposable razors to its members along with other perks? Could you keep your own razor in your locker instead? Do you abuse the free towel priv-ilege, taking three towels to dry yourself, where at home you would have used one? Every towel washed and dried takes more energy. Do you shower for 20 minutes because it's included in your fees? Water heating costs energy, which degrades the environment.

Turn on the dishwasher, washing machine or clothes dryer after 11 P.M.

Remembering that we build our electrical plants to cover peak usage periods, it comes as no surprise to learn that we're doing ourselves and the environment a favor if we run our heavy duty appliances in the off-peak hours.

There are bonus benefits to this plan. Large appliances generate considerable heat. In the summertime, our house can handle the extra heat more easily at night than during the hot daytime. In the wintertime, we turn the furnace down at night, so the extra heat from the appliances after 11 P.M. is a bonus, helping to warm our house.

We find it convenient to start the large appliances just as we're shutting down the house for the night. It's really become a habit at our house, and one that we rather enjoy. If you can't manage that timetable, at least try not to use your big appliances during the highest peaks of energy usage, between 5 and 7 P.M. Furthermore, since these big machines all make noise, it's nice to have them going at night when we're off in the bedroom where we can't hear them.

Use the energy efficiency cycle on your dishwasher and let the dishes air dry.

After you've turned on the dishwasher late at night, do all your bedtime ablutions. Then when you're all ready, make one last trip to the kitchen to cancel the drying cycle on the dishwasher, and let the dishes air dry. On an older dishwasher which doesn't have an energy efficient cycle, simply push the "light wash" button to wash the dishes, and the "cancel" button to stop it from drying the dishes. The light wash cycle washes just as well as the regular wash cycle, and takes much less time, less hot water and less energy. If you always use the regular cycle, try the light wash cycle. You may be surprised at what a good job it does.

It's not necessary to let the machine dry the dishes. Just open the dishwasher door and slide out the racks for a few minutes. The dishes are usually hot enough after washing to dry very quickly in the air.

Use the smallest possible appliance for any cooking job.

When we used to have woodstoves burning away all day, it didn't much matter what we cooked in. The stove was always hot anyway. Now that we're trying to save energy, we realize that it's not such a great idea to warm up the whole oven just to reheat a piece of pizza.

For example, it takes a lot of energy to heat the stovetop element, the pot or frying pan, as well as the air around it. For a small cooking job, a more efficient appliance is the electric frying pan or the toaster oven. My sister Elizabeth even makes cakes in her toaster oven. Saves heating up that big stove oven.

Use a pressure cooker when possible, instead of conventional pots.

You might already be aware that the microwave oven is a good energy saver. It heats only the food, not the oven walls, the air inside the oven, or the food container. It can cut down on the time to prepare meals, as well as saving up to 50% of the energy used by a regular oven. However, the microwave is less efficient than the stovetop for many things. If boiling more than a cup of water, use the stove element, or even better, the electric kettle.

There's a much cheaper alternative to the microwave: the pressure cooker. Elizabeth calls it the poor person's microwave. For the not yet converted, the most frequent worry is whether they're safe, and whether they will explode. Yes they're safe. No, they won't explode. I've been using my pressure cooker for over 10 years, and am a true believer. Yes they save time—I can cook a stew in 10 minutes, *coq au vin* in 12 minutes. And yes, they do save energy.

Bonus: more vitamins per vegetable, delicious one-pot meals, last-minute suppers in a flash.

If you're just starting out, you'll get an instruction and recipe booklet with your new pressure cooker. Read it carefully, and you'll soon be cooking happily.

Defrost foods in the refrigerator.

The frozen foods will help keep the rest of the refrigerator cold, reducing the energy needed to run it. And you'll save the energy you would have spent defrosting in the microwave or oven. A frozen casserole put in the refrigerator in the morning will be ready to heat by suppertime.

Decide what you want before opening the fridge door.

Do you or your children go window shopping with the fridge door open? Sometimes we want a snack, but we don't know exactly what we want, so we open the door and browse. Every time we leave the door open, cold air escapes into the warmer room, and the fridge has to work harder to maintain its temperature.

I've been trying a little memory game that helps save refrigerator energy. Whenever I want a snack, before I open the fridge door I tell myself that I already know everything that's in there. I picture the cheese, the milk, the carrots, celery, apples, or leftover casserole. See if the kids can name every snack food that's in the fridge without looking. They might surprise you with some you didn't think of. The reward, of course, is they get to eat the snack.

Choose manual gadgets over electrical.

How many electrical gadgets have been invented to replace perfectly good manual gadgets? Some that spring immediately to mind are can openers, carving knives, toothbrushes, and pencil sharpeners. Dozens more tempt consumers with promises of faster, easier, or more convenient service. While these may seem like trivial energy users, they are part of a pattern of our profligate waste of energy. They are symbols of a society that has for too long believed that we can have whatever we want, however large or small, without worrying about the cost.

If we reduce our dependence on electrical gadgets, and resist the temptation to buy them for ourselves or for gifts, we'll reduce our demand on the electrical supply, as well as cutting back on over-buying of over-packaged goods. A good quality, heavy duty, hand-operated can opener (the kind with coated handles and a heavy turning mechanism) makes a wonderful gift to a person setting up housekeeping. I've had the same one for about 15 years, and it still works perfectly.

Praise and reward children for turning off lights.

How can we teach the kids to turn off the lights? Bribe them. Let them calculate their bonuses from your utility bills. Some bills have an item for each billing period comparing last year's usage to this year's. That makes it easy to tell if you're getting more energy efficient in your house or apartment. If you don't have a comparison box on your bill, get the kids to keep track of the kilowatt hours used each month (printed on your bill), and let them be the family watchdog for energy efficiency.

For younger kids, post a sheet of paper on the fridge with everyone's name on it. Whoever turns off someone else's lights gets a point. Or whoever remembers to turn off his or her own light gets a point. Award prizes (pocket change or treats) for points. Be lavish in your praise of kids who remember to turn off lights. Caroline is starting to remember on her own, and feels rightly proud when we notice her efforts. Undoubtedly, it won't be long before she's reminding us. For more energy-saving tips, see Chapter 9: Home Renovations and Decor.

What To Do In The Winter

Keep unoccupied rooms at 15° C or 60° F during cold weather.

Turn down the heat controls for the rooms not in use. This is especially important if you have a spare room that's seldom used. It can mean closing the duct cover, turning down the radiators, or turning down the electrical thermostat for that room. But beware turning the heat right off, especially in very cold weather. You could lose heat from heated rooms through the walls, and could have a problem with mold, mildew or condensation in the unheated room. And it will cost more to get the house back to a comfortable temperature.

Shut off your kitchen hot air duct.

If you'll be in the kitchen baking for several hours, your oven will put out plenty of heat to keep the kitchen warm. If you have a warm air duct in the kitchen, close it and let the furnace spend its energy elsewhere.

Turn down the heat at least 5° C or 8° F if you're going to be away for eight or nine hours.

If you're going away for a day or more, you can turn the heat down even more. It's pointless to have the furnace keeping the house at 72° F when there's nobody home. The first one in the door can turn it back up again.

Set nighttime temperature at 16° to 18° C or 61° to 64° F.

We don't need to warm our kitchen tables and our living room couches while we're snug in bed. Save at least 5% to 10% on the heating bill. At our house, we use the lower setting and an extra blanket. A wonderful convenience is a programmable thermostat that will turn your furnace down at night and up in the morning. Available at any hardware or home center the thermostat is easily installed by the home handyperson.

Many can be programmed to turn the heat up and down several times in 24 hours, a convenience if the house is unoccupied during the daytime.

For every degree you set your thermostat above 20° C or 68° F, your heating costs increase by 5%.

Don't forget to reprogram your thermostat if you change to or from daylight savings time.

Leave your bathwater in the tub overnight.

Do you take a hot bath late at night? All that hot water, which you've paid to heat, could be put to a second use. Why heat the sewer pipes? Leave the bathroom door open, and get free heat for your house while you sleep. The warm moist air is nice in the dry winter too. But you'd better be first up in the morning to pull the plug, or someone's bound to complain.

Cover cracks under the door to keep the heat in and the cold out.

What's a "stuffed snake"? It's just what it sounds like—a snake-shaped tube of cotton or other fabric, stuffed with sand or wool or cotton batting. Why would you want one? They're a great temporary insulator for those cracks under the door where the breeze blows through. Find them at craft shows or make your own in less than an hour with leftover fabric scraps. If you can't find or make a stuffed snake, just pull the hall rug up over the crack in the door every night before bedtime. And of course, weatherstrip that door as soon as possible.

Cover windows at night, uncover them in the daytime.

Close your drapes at night. You can buy energy-saving blinds and drapes, but any heavy fabric will be helpful in keeping in the heat. Before we could afford drapes, we used to hang a blanket in front of our bedroom window on winter nights. It's remarkable how efficient a blanket can be. The sun will help to warm up your house on cold winter days. Don't shut it out.

Turn off your attic power ventilator.

Some newer homes are equipped with a power ventilator in the attic. It's meant for summer use only. In the winter, it will draw warm air up from the house into the attic, increasing your heating and energy costs. It could also cause frost buildup and damage in the attic from the moist air which it draws in.

> **Use a hot water bottle, and turn down the heat for more energy and dollar savings.**

In the house where I grew up, we had one hot air duct out in the hall serving three bedrooms. Winter nights are extremely cold on the Canadian prairies. My smart mother kept her four kids warm at night with hot water bottles. But being a woman ahead of her time (and taking a lesson from the past), she didn't go out and buy four rubber hot water bottles. She reused what she already had—glass sealer jars. We may have had the odd leak in the bed, but she remedied that by putting a piece of plastic over the mouth of the jar before putting on the rubber ring and screw top.

Caroline loves to cuddle up to her hot water bottle on cold nights. I think it helps her fall asleep more quickly too. And don't forget to wear your socks.

What To Do In The Summer

> **Turn up the central air conditioner temperature at least 5° C or 8° F if you're going to be away for eight or nine hours.**

Since heating our homes uses the most energy in winter, it stands to reason that cooling our homes uses the most electricity in the summer. It makes sense to take the heating tips for winter and throw them into reverse for the summer. If you're going away for a day or more, turn the central air-conditioning right off.

Turn off room air conditioners if you'll be out of the room for more than five minutes.

Room air conditioners don't do much good if you're not in the room, and they do draw a lot of electricity—anywhere from 750 to 1050 watts, depending on the size. We would never leave a lightbulb of that wattage burning while we left the room. Since we get almost instant gratification from turning on a room air conditioner, there's little advantage to leaving them running while we're out.

Keep your curtains or blinds closed during the hottest part of the day.

Just as the drapes kept out the cold in the winter, they'll keep out the heat in the summer. And, yes, close your windows behind your blinds too. Even though you might think you'll get "fresh air" from outside to cool you, what you'll really get is hot air from outside if you leave your windows open during the day. However, if you have no curtain, blind, or window covering, you may wish to leave the window open a bit to let out the hot air from inside—remember the greenhouse effect.

Set your light timer to turn lights on later at night and off earlier in the morning.

If you have an automatic light timer that turns your security lights on at a certain time of day, remember that days are longer in the summer time. A light-sensing device will make the change automatically for you. If you're investing in security lighting, buy a unit that turns on automatically at dusk and off automatically at dawn. You'll save money and energy.

Turn off your fan if you'll be out of the room for more than five minutes.

If you have a small room fan, it won't be of much use to you when you aren't in the room. But a portable fan is a wonderful, low cost alternative to an air conditioner. A room air conditioner can use up to 20 times the energy of a portable fan. A single portable fan sees our family of three through the worst summer heat waves. We usually just use it at night when we're going to sleep, or in my third floor office when I'm working there.

Turn off lights, television and appliances when not in use.

Don't leave your television, electric appliances, computers, or lights on in another room. Not only will you be wasting energy, but all appliances, and especially lights, generate heat. Over the course of a hot day, shutting them off can help a lot.

Open your bathroom window or turn on the exhaust fan after your shower.

Remember all that warm moist air that we wanted to keep in the house after a late night winter bath? Well, we don't want it any more. Summer air is warm and moist already, so we don't need more of it in the house. Be sure to close the window and shut off the fan after the warm moist air has dissipated.

Iron outdoors on your deck, patio, or balcony instead of inside the house or apartment.

My mother always irons in the basement, because it's cooler, and I can certainly recommend that idea. But my basement isn't a place for happy ironing (nowhere is, in my opinion—I'd rather have more wrinkles and less ironing). If I have a lot of ironing to do, I prefer to put myself in the best environment possible. Enjoy nature, and leave the heat outside in the summer. It makes a big difference.

Put the sunshine and warm air to work for you.

Take advantage of the sunshine and warm temperatures for the jobs that usually require chemicals or heat. Whiten clothes by spreading them out in the sunshine—it really does work. I've done it dozens of times with good to spectacular results. Use a clothesline to dry clothes. Dry your hair in the sunshine (does it really make your hair shinier?—I must be imagining that). Hang bunches of summer herbs to dry in the sun. Fill the kids' wading pool early in the day in a spot where the sun will warm it now, but trees will shade it later when it's splash time.

If You Have More Time

Insulate your hot water heater.

Hot water heaters use more energy than lights and appliances combined. There are a couple of ways to cut back. You can buy a water heater "blanket," a padding of insulation made to fit your heater. They cost about $20 at a hardware store. Or wrap your own insulation around the tank with the vapor barrier to the outside, and tape it with duct tape. This job can be done in less than an hour, and will give savings all year.

Plant a tree today for future generations.

Trees are worth hundreds of thousands of BTU's cooling. See Chapter 12 to find out about the wonderful tree that's my substitute air conditioner. A mature tree will cool yards, gardens and houses by up to 15° in the hottest part of the day. Trees that are mature today may be dying or dead at the same time we are. Let's plant more.

> In North America, every person consumes the energy equivalent of 22 barrels of oil at home, every year. There's plenty of room for you to conserve comfortably with just a small effort.

7

GREENING UP
AT WORK

Share The Rewards From Top To Bottom

The workplace these days may be blatantly hazardous, using toxic chemicals or polluting the air or water. Or it may be disguised as a normal, ordinary, functioning office, when in reality, it's chewing up natural resources and costing the environment dearly.

The purpose of this book is to offer suggestions for individual action on a small scale. The nice thing about small actions within a company is that you quickly change the habits of a large group of people. You will find many ideas in this chapter that are within your power to accomplish easily. All of them can have a large, positive effect on the environment.

You can also make yourself popular at the same time. You could become popular with the boss (if you aren't the boss) by saving money for the company through environment-friendly procedures, most of which are economical measures the company should be following anyway. Improving working conditions and being a leader in a movement to benefit everyone will make you popular with your co-workers.

If you're the boss, insist that the workplace be environmentally friendly. Post the rules and regulations and make sure they're obeyed. You'll likely get a lot of co-operation and enthusiastic support if you take the time to tell the employees the reasons for "greening" the company, and ask for their suggestions. Everyone benefits—the company, the workers, the community and the rest of the world.

Even if you're not management, your influence will be felt as you do your own small part in the greening of your workplace. When you do something differently from the normal procedure, explain to your co-workers and your manager the reason for the change, and the many benefits to them and the company. Promote environmental awareness in your company newsletter. When even one or two small changes start to make a difference to the planet, everyone will begin to get enthusiastic about her or his own contributions, and become more conscientious.

What To Do

Photocopy or print on both sides of the page.

Does your office use a lot of paper? How many filing cabinets does it take to hold the paper that your office generates? How many shelves does it take to hold a year's supply of paper? How many trees does your office use in a year?

Cut your paper expense in half at one stroke, and do our forests a favor by photocopying on both sides. This goes for all reports, documents, scripts, data sheets, even business letters. If you think people will be confused or upset with this change, on the front page put "(over)" in the lower right hand corner. If any explanation is needed, try "We're printing on both sides of the page to save money and save our forests. Ask us for details!" I'm willing to bet there will be no complaints, and lots of support.

If you happen to be the only one who has figured out how to make photocopies on both sides so that the printing comes out right side up, take two minutes to write a clear instruction sheet, with pictures, and post it above your photocopier. Encourage people to ask you for help.

At my publisher's office, all ruined copies go into the bottom tray of the copying machine. Then they select "Tray 2" when they have interoffice memos or other internal documents to copy.

Reduce the number of copies of any given document.

How many copies do you really need? Is there one for the customer, one for the secretary's files, one for accounting, one for purchasing, one for the boss, one for "just-in-case"? Some companies spend two days printing end-of-month reports because executives don't want to share. See who files copies without ever referring to them again. If it's on computer disk, do they really need a hard copy? Could they use another department's copy?

copy? Could they use another department's copy?

If you're making copies of a document for a meeting, try to determine the exact number of people who will need them. If they don't need to take copies away with them, perhaps the information could go on a blackboard, or on one or two paper copies to be shared around the table. Circulate one copy to all departments before or after the meeting. Avoid extras.

You might want to make a list of all reports, and ask people to mark the ones they use. You'll be surprised how many people accept long reports they never read, just because the reports arrive.

Reuse paper until both sides are used.

Leave a box beside the office copy machine for "ruined" copies. Everyone can use these "good one side" pages for scrap paper, rough drafts, telephone messages, or steno pads. If you're feeling especially energetic or enthusiastic, take a handful of these pages, cut them in halves or quarters, and staple the corners together to make message pads or scrap paper pads for everyone in the office. Soon they'll be doing their own pads. You might want to check with the office manager to learn how much the company previously spent on note pads. Then keep track of the savings and let your boss know the bottom-line results of your environmental efforts.

Recycle office paper.

Now that you've printed on both sides, and reused the second side of every piece of paper, have you still got piles of paper going in the garbage? There are numerous companies who will help you get started in paper recycling. Check the yellow pages of your phone book under "Recycling," or call any paper companies in your community. Domtar has a recycling program that provides large plastic bins for your company. They pick up the paper on a regular basis. Many other companies will gladly pick up your paper waste. It won't cost anything, and all it takes is one person who is willing to make a phone call.

Contact a bicycle courier company.

Do you send packages by courier? Many cities now boast bicycle courier services. It seems a terrific waste of resources for a car to burn fuel all the way across the city and back again to deliver a single envelope. And as the traffic problem becomes worse in city after city, bicycles are often the fastest method available. If your company has its own delivery person, suggest that shorter trips can be made by bicycle—but just for the sake of a little environmental altruism, pay them the same mileage rates as if they took their car.

Recycle corrugated cardboard.

Do you receive boxes and boxes of shipments? Check with the same paper companies about recycling. Some municipal garbage dumps are now refusing to accept corrugated cardboard because of the bulk, and because it is so readily recycled. Reusing is even better than recycling. Reuse wherever possible. Will your suppliers take back their empty boxes to reuse? You'll never know if you don't ask them. You can save them a lot of money, savings which they should be willing to pass on to you. Flattened boxes from previous shipments could be returned to the company on the same truck that brings the full boxes.

Take your own ceramic or china coffee cup to work.

Coffee break. What's the system at your place of work? Do you have a coffee center? A cart that comes to the door? A truck on the street? Or do you run to the restaurant next door? How do you take your coffee—in plastic, paper or ceramic? I hate the thought of the millions of white plastic foam coffee cups that go in the garbage every day, not to mention all the little plastic stir sticks that go with them.

Don't be fooled by plastic foam cups or plates being touted as environment-friendly. These cups have been produced without using chlorofluorocarbons, which are damaging to our ozone layer. Instead they use another blowing agent, which is considerably less harmful. But what they don't say is what will happen to those millions of coffee cups when they're thrown in the garbage, after less than 10 minutes use. Like the rest of our garbage, they go to landfill sites or incinerators, where they either sit for a thousand years without decomposing, or produce toxic fumes when they burn. What's the alternative? Paper cups are somewhat better, since at least they're biodegradable, and made from a renewable resource. But it's silly to chop down forests just for coffee cups. A ceramic or china cup will last for years.

Grab your cup from your desk when you head out for coffee. If you have a regular cart or truck that comes around, get to know the server or owner. Decide between the two of you what's the fill up mark on your own cup. Who knows? You might start a trend when other customers look on in envy at your permanent coffee cup. Maybe the owner will give you a free cup of coffee at the end of the year, for all the disposable cups you've saved.

Of course, if you have your own coffee center at work, every worker should have her or his own cup. Guest cups should be washed by the person entertaining the guest—all employees and bosses wash their own.

Make a sign for the inside of the front door that says: "Last one out shut off the lights."

This simple trick will produce good results. Lighting can account for a huge portion of company operating expenses. Check your latest electrical bill. Don't forget those washroom lights and individual desk lamps. But don't stop at lighting. Turn off computers, photocopiers, coffee machines, heaters, fans, typewriters, printers and any machinery. And of course, turn down the heat. Gain the support and co-operation of everyone at work by offering a share in the savings of a reduced electrical bill.

Offer a small discount or rebate to customers who bring their own bags and boxes.

You can also provide or sell reusable cloth bags, like the kind that are popular in Europe. Not only does the environment win, but you'll save money and so will your customers. Be sure to thank them effusively for bringing their own bags or boxes. That will encourage them as well as other customers who are within earshot.

If You Have More Time

Investigate alternative cleaning materials.

What cleaning materials are used in your workplace? Is your company still using industrial-strength cleaners and solvents? Is it really necessary? Keep a box of baking soda in the bathroom and label it "Sink Cleaner" in big black letters. Buy your boss a bottle of environment-friendly dish cleaner for a birthday present. Check Chapter 3 for other ideas. Be sure to enlist the support of your office cleaners or janitors. They may be happy to switch to products which are less toxic, since they work with them constantly.

Find the facts on health hazards in your industry or business.

Do you work in a hazardous industry? Is your health or the health of fellow workers threatened by occupational practices? Get information to help you educate your company or, if necessary, force them to provide protection for you. The Ontario Environment Network's Resource Book ($6) lists pamphlets such as: "A Worker's Guide to Solvent Hazards," "How to Protect Your Health at Work," "Smoke in the Workplace Action Kit," "The Magic and Deadly Dust: Asbestos and Your Health" and "Your Office Job Can Make You Sick." Their address is in Chapter 16. Or write for A *Citizen's Toxic Waste Audit Manual* ($5 donation requested) from WAM, Greenpeace Toxic Campaign, 1017 W. Jackson Blvd., Chicago, Illinois 60607.

Weatherproof your workplace.

Just like at home, you'll save money if you save energy. That means weatherstripping doors and windows, insulating walls and ceilings, insulating water heaters and water pipes, and turning down the heat at night. The bigger your workplace, the more you'll save, and the sooner you'll make back your investment.

Get everyone involved in thinking green.

How about a prize for the best "green-up the workplace" suggestion? Offer a potted plant as an appropriate prize for the best idea. You don't have to be the boss to get this idea going. You just have to enlist the boss's support. You'll be amazed at what the rest of the folks will think of without any prompting at all. And you'll suddenly discover that you don't have to save the world all by yourself. That's a truly uplifting feeling.

If you own a food shop or restaurant, try some organically grown produce or meat.

Customer response has consistently supported organic agriculture—even when it means paying higher prices.
Support local growers—fewer trucking miles means less air pollution. When we eat imported foods, we add costs to the environment, as well as to the food itself.

Every truckload of imported food burns gasoline all the way from origin to destination. Buying local produce is a step towards cleaner air, a step towards the reduction of the greenhouse effect and a step towards saving our agricultural lands.

Turn your company's garbage into gold.

Many waste materials from one industry are in demand from another industry. Get together with another company to reduce waste from both companies, and you can both save money. Pollution Probe has published an excellent book on this topic called *Profit from Pollution Prevention*. It's available from Pollution Probe, 12 Madison Avenue, Toronto, Ontario, M5R 2S1, for $25. Or use your credit card and call them at (416) 926-1907. Expected this year is the first of a series of smaller books updating the same topic.

Plant trees on company property.

Is there room for a few trees on your industrial worksite, in your company parking lot, or by your front door? Every new tree planted helps reduce our atmospheric carbon dioxide, offset the greenhouse effect, filter pollutants, and stave off the concrete canyons.

With a few quick phone calls, you might even be able to get someone else to plant them for you. Try your local Boy Scouts, Girl Guides, church group or environmental group. For a donation, someone will certainly do the job. Many cities or municipalities will plant trees on their own property which adjoins yours. Check with city hall. If you can't get someone else to do it, why not organize the employees? The annual company picnic is a perfect time to have a tree planting—just before the baseball game. Remember Arbor Day? We used to observe it by planting trees. What a wonderful tradition to revive. I bet you could even get some free publicity for it.

Help educate your company about the environment.

Environmental films, videos and speakers are available for your company meetings. Check with your local environmental groups to see if they have any speakers. If not, they'll certainly have suggestions. Presentations can be geared to your specific company, or you can have a general awareness presentation. Invite everyone from the president to the caretakers. Try to organize a presentation for lunch hour or even on company time.

Think globally, act locally—a well-worn environmental slogan, but one that works. Get started at your desk. You can change a large community in just a few minutes.

8

SCHOOL
DAYS

Those Good Old Golden Rule Days

Aside from our parents and families, our schools may
have the most lasting influence on our lives. Sometimes
I remember my school days with gratitude, and other
times with dismay.

Now that I have a daughter in school, I'm reliving some
of my past. When I see her at school, I try to imagine
how her perception of these days will filter through to her
when she is an adult. What I want most for Caroline is
for her to have happy memories, lasting friendships, and
an unbridled enthusiasm for learning.

My daughter's school year lasts about 186 days. She
spends about seven hours a day at school, including lunch
time. That means over 1300 hours in school every year.
That's a huge chunk of time in a child's life. The environmental influences and the environmental awareness
of her school will form an indelible impression on her for
life.

Although we can't manufacture happiness for our children, we can do our utmost to provide the setting and the conditions to foster the security and sense of one's place in the community necessary to finding happiness. We have the right to know that when we kiss them good-bye in the morning, they're going off to a school where they will be free from physical danger. Equally important, we want them to be part of a community that not only protects them, but actively enhances the world around them.

That means making environmental awareness a part of every school lesson and every school action. If you can foster in the teachers and principals the knowledge that every subject is affected by the environment and every action affects the environment, you'll go a long way to infusing a love for the planet in the children. Once that lesson sinks in, we're on our way to a greener future for the Earth.

We owe it to our children to give them the greenest, healthiest Earth possible, and also to give them the tools, the knowledge and the wisdom to carry on the process in the future.

What To Do

Stand up at the next PTA meeting and propose an Environment Committee for your school.

One of the best ways to ensure an environment-friendly school is by getting everyone involved—principals, teachers, students, caretakers, teaching assistants and parents.

The quickest and easiest way to do that is to start a committee. Don't groan. Let me tell you a brief bit about what happened at my daughter's school.

I started by going to the PTA meeting, and asking whether they were interested in environmental issues. I explained a bit about some of the issues that affect our children directly, and the planet as a whole, and some of the things that we could do to encourage environmental action at school. I was overwhelmed by the enthusiasm and energetic response.

I volunteered to head an Environmental Issues Committee. I thought it would take a year or more to get the ball rolling, but by our second meeting, we had so many things happening that I barely had to do anything. The principal and teachers not only grasped the issues immediately, but were starting to take action on their own. Suddenly I found I didn't have to save the world single-handedly. Was I ever relieved!

Use any ideas you find in this book to make your job easier. If you get even one teacher or one other parent on your side (and believe me, you'll get lots more than that), the job will become a joy. They'll accomplish so much more than you could ever do on your own.

Pack a garbageless lunch.

Have you ever seen a school garbage can after lunch time? Overflowing, isn't it? There's one simple rule for environmentally friendly school lunches. Don't include any throw-away junk. A few small purchases should last your child for many years. A plastic lunch box or a cloth lunch bag, and an insulated or vacuum bottle (thermos) will eliminate plastic bags, paper bags, individual drink boxes and plastic wrap.

At our school, I discussed this concept very briefly in our first environmental committee meeting. A grade two teacher took it back to her class. By our second meeting, she had 100% participation in garbageless lunches. She didn't even send a letter home. She simply explained the concept to the grade twos, and they enlisted their parents' co-operation. We were all thrilled.

Children sometimes complain that they want their lunches to conform to some ideal so they won't seem different. If we can catch them early enough in their school years, we can ensure that everyone in the school adheres to a green standard. Even in later school years we can teach them the concept, and they'll adopt it quickly. Caroline has a red cloth lunch bag with a Velcro closer. We wash out plastic milk bags to hold her sandwiches. After just a few reminder notes tucked into her bag, she learned to bring home any plastic bags in her cloth lunch bag for reuse. She's proud of her environmentally friendly lunches. We're proud of her.

Before you buy all your new supplies in September, take a few minutes to look for all the leftover supplies from June.

School supplies—do we really need everything new every September? Oh yes, I do remember that wonderful feeling of walking to school the first day with new unsharpened pencils, unblemished notebooks, clean rulers and erasers. But what happened to last year's leftovers? How did we learn to discard perfectly good clothes, cars, and household equipment on a yearly basis? Could it have started from those early school day influences? Does starting with new material goods mean a fresh start in thoughts, ideas, philosophies and ideals? Or does it simply distract us from the fact that we're leading the same old lives with the same old ideas?

Caroline seems to have no objection to starting her school year with supplies which we scrounge from her desk and around the house. Perhaps that's because we've never made a ritual of purchasing everything new every September. We're trying to instill in her (and ourselves) the idea that new does not necessarily mean better. The same philosophy can be applied to all school purchases. Textbooks, desks, chairs, televisions and film projectors all have an impact on our environment in their manufacture, sale and eventual disposal. All are items which can be repaired. Think twice before discarding something just because it's old.

Children pick up ideas from very ordinary expressions. So let's remember to praise some "good old things" because they're sturdy, reliable, accurate, useful, familiar, as well as exclaiming over "brand new things."

Take some used photocopy paper from the office to the school.

Is there any way to cut down on paper use in an institution that thrives on paper? We've already discussed ideas elsewhere in this book, such as double-sided photocopying, making scrap or art paper from paper printed on one side only, and recycling fine paper. Reducing is always the best option. Our vice-principal thought of an idea which reduced our school's use of paper considerably. She often receives notices of various meetings, classes, projects, or information that would be useful to some or all of the teachers.

She had no way of knowing which information would be of interest to which teachers. Her previous practice had been to photocopy every notice and give a copy to every teacher. Thinking of saving paper, she decided to make a loose-leaf binder for the information, and leave it in the staff room where all the teachers could check it regularly. See what I mean? All you have to do is plant a seed, and the rest of the people will take it from there.

If You Have More Time

Start an anti-litter campaign.

Don't litter—one of the earliest environmental lessons. Is your schoolyard a mess? Do the caretakers have to pick up the junk from classrooms, hallways, gym or lunchroom? A school campaign to be litter-free will pay off in several ways. The school will look better, everyone will feel better, and the students will begin to grasp the concept of a shared society.

When we discard junk carelessly, we're polluting the world that we all share. Will the child who never learns not to throw plastic candy wrappers on the ground become the adult who dumps toxic pollutants into our rivers? If those adults who now spew toxic fumes into our atmosphere had learned not to litter when they were children, would they think twice before behaving so badly toward our planet and its inhabitants?

If your child learns the basic anti-littering lesson, she

or he will be on the way toward becoming an environmentally responsible adult.

Brainstorm with students and teachers on environmental projects.

Class projects and outings are a perfect opportunity to learn about the environment. Creative teachers and students will come up with a lot more ideas than these few. Clean up the environment, from the lunchroom to the schoolyard to the river and beyond. Will the school board let the kids caulk the windows? Build bird houses? Plant trees, flowers, ground cover around the school yard? Make a compost pile? Make a worm composter (kids are crazy about them) for school lunch leftovers (see Chapter 11). Sew cloth lunch bags in the school colors. All these activities can be geared to lower or higher grades.

Make environmental studies a part of every subject taught in your school.

Try to get teachers from every subject involved. How can you relate mathematics to the environment? Try some statistical fun. If every student makes four pounds of garbage every day, how much garbage does the class produce in a year? If you save two gallons of water by turning off the tap every time you brush your teeth, how much water can be saved if the whole school participates?

• **History and the environment:** When did air pollution become a serious problem? In 1970, when we first started hearing about it? In 1860, with the industrial revolution? In 1492, when burning wood and coal was the only way to produce heat?

• **Geography and the environment:** What's the effect of the destruction of the ozone layer?

• **Social Studies and the environment:** How much are industrialized countries to blame for wreaking havoc on the environment? Should developing countries pay the price of fewer conveniences and a lower standard of living in order to alleviate the damage we've done to the environment?

• **Reading, writing and the environment:** Numerous books are available on a wide range of environmental and nature study topics. Writing assignments can include essays on environmental problems and solutions.

Get the students involved in making the environment a top priority for your school.

How about a contest to promote environmental awareness? Some ideas: an environmental poster contest, a fund raiser for school trees, a flower planting contest, an environmental slogan contest. Let the kids make it up, and be the judges themselves. Let teams work together to vie for "green" prizes. Pretty soon the students will be telling you to clean up your act.

Request a non-toxic environment in your school.

Time to paint the school? Every year some students are forced from their classrooms, because of severe reactions to fresh paint. Ask the principal to use non-toxic paints. Chapter 9 tells more about what to choose. With non-toxic paints, kids can even contribute to school decor by doing a hallway wall mural themselves. Another idea for an environmental design contest.

Choose environment-friendly cleaning supplies for your school.

What goes down the drain at your school? We don't usually see what our caretakers use after hours, so we don't usually care. But schools are big users of cleaning supplies. If we're learning to choose alternatives at home because we care about what goes into our sewers and ultimately our lakes and rivers, then we should care about the cleaning materials used in our schools.

To do this, you'll have to enlist the co-operation of your caretakers, and probably the school board. This calls for the principal's involvement, and some information and education about the reasons we need to be concerned. If you need some ammunition, try Chapter 3.

Search out help from the school board, the community and environmental groups.

Remember that other groups are involved in the environmental movement too, and you can often find help in the most obvious places. Check with your school board to see if they have an environmental consultant, or an environmental studies kit. Phone your local environmental groups to see if they have speakers or films suitable for the school. Check your teachers' publications for environmental activities ideas. Soon you'll see the movement taking off under its own steam.

Our children are our most precious resource. They deserve every chance to survive and conquer whatever mess we've made of the Earth. Let's not send our kids off to toxic schools. Let's make our schools reflect our beliefs in the future of the planet.

9

HOME RENOVATIONS AND DECOR

What I Learned From Dad

When I was about five years old, Dad set up his saw horses and saw outside our back door to work on one of his dozens of building projects around the house. I have a vivid memory of gathering the little ends of wood and making numerous variations on my own housing project. At age eleven, I began drawing detailed housing plans to amuse myself on rainy days.

Around the same time, Dad taught me how to straighten used nails. He never threw away nails from a renovation project, but it was a tedious job straightening them for use again. My brother David and I were taught to pull bent nails from boards, hammer them flat on the concrete sidewalk and put the straightened nails into a box. As teenagers, we helped shingle the roof at our cottage, and assisted at the other end of the saw on various projects.

Later, Barry taught me more about carpentry, and

when I started renovating our own house, I learned plumbing, wiring, plastering and painting from books, and trial and error. Building and renovating seems to be in my blood. But I think Dad taught me the most valuable and lasting lesson: save old materials for new projects, and never buy new what you already have.

Dad still has stacks of used lumber, insulation, electrical switches, plumbing pipes and other useful building materials from dismantled buildings and projects. When he starts a new project, he has a great stash of ready-made resources. He often laughs at himself, and jokes that for some certain project he actually *bought* some nails.

Yet there's a sense of pride in the money and materials saved, not squandered. And Dad is one who really knows how to make a silk purse from a sow's ear. Last summer, he built us an excellent brand new window screen from some dusty old pieces of wood that he found in our woodshed.

In North America, we tend to treat our resources as if they were limitless. Any garbage day, you can see useful and expensive items sitting out at the curb, waiting to go to our landfill sites. People throw out everything from tables and toilets to mirrors and doors. Even the kitchen sink. All these items cost our environment in their manufacture, their transportation and their eventual disposal. Meanwhile, other people scrimp and save to try to afford the least expensive of new items—often of lesser quality than those discarded. Then these items, too, eventually wind up in the garbage, and it's back to the store for more.

Since Barry and I are a bit nomadic, I've renovated several houses over the years. I know the enormous quantity of materials that go into and out of a building or renovation project. I realize that it's almost impossible to improve a house or apartment without wasting some materials. But if we start thinking about alternatives as we renovate and decorate, we can drastically improve both our own quality of life and health, and the impact we make on our local and global environment.

What To Do

Install a water-saving showerhead.

It gives you just as good a shower while using 60% to 70% less water. You don't need any special tools to install one. Your old showerhead just unscrews, and the new one screws on. Water saving showerheads cost anywhere from $15 to $65. But remember that the water heater accounts for a huge portion of your energy costs. Your water heating bills will be lower, as well as your water meter bills.

If everyone installed water-saving toilets and showerheads, we could reduce domestic water consumption by up to 75%. We might never have to build another water treatment facility.

Choose one higher-watt bulb rather than two lower-watt bulbs.

A single 100-watt bulb will give the same amount of light as two 60-watt bulbs. But it uses up to 15% less energy. In applications where you have the option, such as bedroom or kitchen ceilings, garages or basements, choose the single bulb fixture over the double bulb fixture.

Install and use dimmer switches on lights.

Turning down the brightness on incandescent lights will save energy, as well as extend the life of your lightbulbs. Contrary to popular belief, the extra energy required for full brightness does not somehow leak out of the switch when the lights are dimmed. The flow of energy is, in fact, reduced.

Reuse, reuse, reuse!

Think of the huge dumpsters sitting outside home building or renovation projects, all full of discarded housing materials waiting to go to our overflowing landfill sites. My local Home Builders' Association estimates that they waste over two and one-half tons of building materials, including drywall, masonry, tiles and plastics, for every house they build. Much of this material could be sorted and reused, and that is just what the Association is proposing to do. On the smaller scale of individual homeowners, we can do no less.

• Carpenting: How many things can you do with a piece of old carpeting? Put it in the basement laundry room or rec room. Cut out the worn spots and carpet the back porch. Put it in the trunk of the car to keep the trunk clean. Hang it on the walls of the furnace room as insulation. Make your cat a scratching post.

• Mirrors: Cut off the chipped corners and build a new frame. Or build a deep frame that covers the chipped corners. If the mirror has beveled glass, take it for resilvering (mirrors without beveling cost much more to resilver than to replace). Hang it in the basement where you don't care about the chips—it will reflect light and make the basement brighter. Cut broken mirror into strips for beautiful homemade wind chimes.

• Gypsum board: What on earth can you do with pieces of old gypsum board? Believe it or not, small quantities can go into your compost. Larger pieces can be saved for patching holes.

• Nails: Straighten them by laying them curved side up on a hard surface such as concrete. Hammer the curve down flat. Roll the nail a few times to see where bumps remain. If you have a little carpenter's helper, show her or him how to straighten nails—it may keep your helper occupied long enough for you to get some work done. Children as young as eight can handle a hammer without hurting themselves, if given proper instruction.

• Lumber: Again, according to my local Home Builders' Association, 10% of all lumber used to build a house is wasted. Good lumber is infinitely reusable. Try to keep it sorted into sizes so you'll always be able to grab the piece you need. I've been known to carry used lumber from house to house with me. And I always like to "inherit" a lumber pile from a previous owner. It all gets used eventually. Even old lath is good for shimming and wedging.

• Appliances: That old fridge moved to the basement can be a godsend when entertaining—or for storing cold drinks in the summer. Keep it unplugged and the door propped open when not in use. The sink or toilet unsuitable for the new ensuite could be perfect for basement or cottage.

• Light fixtures: Extra light in the basement or garage

can improve safety and make unused space usable. That ugly old fixture will be a conversation piece until finally, one day, it becomes an antique, worth lots of money. Old houses seldom have closet lights. Fixture style is of little consequence in a closet. Reuse a small old fixture in the closet rather than buy a new one.

• Doors: Taken from one area of the house, doors will often fit another, and will maintain the unity of decor better than buying new doors. Save them for newly built rooms, closets or cupboards. Old slab doors make great table tops for workshop, quilting or crafts projects, or children's art projects.

• Windows: Some of those beautiful old pine frames are seeing new life as mirror frames—an easy and rewarding home project. Old single-glazed windows are perfect for "interior" windows—to give extra lighting in stairwells or hallways, or as a feature in an interior wall. Stained glass windows, while drafty in exterior walls, make elegant interior pieces of art.

Someone wants your used goods. Find out who it is.

If you can't reuse it yourself, try to give it to someone who can. Numerous charities will accept good used furniture and appliances. Women's hostels may be able to put you in touch with a person starting a new life, completely without funds or furnishings. Daycare centers will often accept couches, chairs or carpeting. Churches will accept donations for their bazaars.

A garage sale once a year will get rid of a lot of your "junk," and will introduce you to neighbors you might not otherwise meet. You might even make a few dollars. If you don't have enough stuff on your own, get together with friends or neighbors for a joint yard sale. At the end of the day, put up a sign: "Free leftovers." Leave the stuff on the lawn and go in the house, and you'll get rid of a great deal of the rest of it.

Remember, unless the item is extremely precious to you, it's better to have someone else use it than to leave it sitting in your basement for years, where it may become musty, broken, outdated, or unfixable. Advertise in the newspaper to sell your used goods or put up a notice in your local supermarket or laundromat.

Recycle scrap metal.

If you've exhausted all avenues for reusing, scrap metal dealers will often take stuff for recycling—old iron pipes, cast iron bathtubs, radiators, grills, stoves, iron bed springs. Depending on the quantity you have, a scrap dealer may charge you to take it away, take it away for free, or pay you for it.

Support garbage pickers.

They play a useful role in keeping down the quantity of our waste going to landfills and incineration. There's a terrific informal recycling system that happens on the street. If you have a large item that may be useful to someone else, which you haven't been able to sell or give away, put it out at the curb well in advance of your garbage pickup. Chances are it may disappear before the garbage truck arrives.

Look for used items before you purchase new.

The more we reuse, the fewer of the world's resources will be wasted manufacturing new stuff and disposing of old stuff. Check your newspaper for used kitchen cabinets, sinks, toilets, carpets, windows, furnishings. You can even buy used lumber and hardware from wreckers. Up until last year, Barry and I had not one single stick of new furniture, after 17 years of housekeeping. Everything we owned, we bought secondhand or "antique." Last year, we broke down and bought a new piece of furniture. I now regret the purchase and have vowed to try not to let it happen again.

Repair rather than replace, wherever possible.

Consider replastering interior walls, rather than tearing them down and putting up new. Old plaster is smooth, heavy and more sound-insulating than new drywall. I hate seeing dumpsters full of old plaster, and stacks of new drywall or gypsum board going in the door. It seems a terrific waste of materials.

Reuse building by-products.

If you must get into putting up new walls, try to reuse the waste materials. The large plastic pails that ready mixed drywall compound comes in are terrific for household and garden use as water pails, paintbrush cleaning pails, or garbage pails. The smaller ones, with their sturdy handles, are great for kids to use at the beach. The heavy paper or plastic bags left over from concrete and sand projects make excellent garbage bags. A few paint stores will accept empty solvent containers for refill. The corrugated cardboard boxes in which new fixtures are shipped should be recycled if your community has such a program.

Rent or buy a heat gun to remove old paint.

Chemical paint stripper is really toxic, and we'd rather not have it flushed down our sewers or put in our landfill sites. A good alternative? You can rent a very efficient heat gun for $11 to $14 per day from any tool rental store. Check your yellow pages under "rentals." If you're working on a large area, buying a heat gun will be cheaper than stripper. With a heat gun, you avoid compounding the problem of adding toxic stripper to toxic paint. With any method of stripping, be sure to have adequate ventilation.

Put stripped paint in old paint cans and take it to the toxic dump site, or wait for the toxic waste pick-up day.

All that old paint that you've stripped off may contain lead. If it goes in the garbage, it could end up leaching lead into the water table. If you have no toxic dump site and no pick-up day, try to discard your stripped paint in a leak proof container. Then badger your politicians to start a toxic disposal program.

Look for natural paint, or, failing that, use latex rather than oil based paint.

When you're ready to paint again, after all that stripping, consider using natural paint, such as ones made with citric oil and linseed oil base, or interior whitewash made of milk casein. It can be mixed with water and tinted with pure plant materials or earthen and non-toxic mineral pigments. These paints come in a beautiful range of lively, transparent, earthy shades of rainbow colors, with no artificial colors. Check your yellow pages under "paint."

It used to be popular wisdom that you should use oil-based or alkyd paint for kitchens and bathrooms. But the solvents in oil-based paint contain toxic chemicals. I've successfully used latex to paint both kitchen and bathroom. Latex or water-based paint now come in glossy as well as flat finishes, and are both durable and washable.

Make sure all your paint gets used.

There's no reason for half-empty cans of paint to go to the dump. Try to use all your paints, even mixing them together if necessary. My sister Elizabeth and I mixed two different grays, a white and a pink (all latex, and all left over from other painting projects) to get a beautiful soft gray which we used when we painted my front hall. Ask your neighbors if they want your leftover paint, or have a yard sale. Donate your paint to a daycare center or a hostel.

If you absolutely can't find a use for your leftover paint,

take it to the toxic waste site. A municipal garbage dump is not the place for paint. Some municipalities now mix together all paint collected from toxic waste sites. The mixed color usually turns out brown. They then give this paint away free to residents. It's proven very popular for picnic tables, sheds and fences.

Keep solvents out of our water systems.

If you do use alkyd paint, don't put solvents down the drain. When you're cleaning brushes or rollers, use an old empty paint can, or a plastic bucket from drywall compound. Soak and scrub your painting tools. Let the solids settle to the bottom, then decant the solvent from the top. Save the solvent for reuse. Any leftover solvent or sludge should go to the toxic waste collection site.

Check the energy rating when buying new large appliances.

If you're buying new appliances, be sure to check the energy efficiency labels. Even if a particular model of appliance costs a little more initially, you'll save in energy costs over the life of the appliance.

Save yourself the cost of a waste disposal system. Compost those potato peels instead.

Think twice before installing a waste disposal system under your kitchen sink. You could be composting all those kitchen peelings, cores, skins, rinds and seeds, in-

ARJORIE LAMB**

stead of putting them down the drain into our sewer
systems. Furthermore, the waste disposal is yet another
electrical gadget, using energy which could be put to
better purposes. See Chapter 11 for composting instructions and tips.

If You Have More Time

Install water-saving toilets.

If you're replacing bathroom fixtures, look for the new
one and one-quarter-gallon flush toilets, as opposed to
the old five-gallon flush. A family of four can save up to
32,000 gallons of water per year. If you are on a water
meter, you could save considerably.

Weatherproof your house.

If you're opening up your walls anyway, be sure to
insulate to the maximum. You'll save in the long run by
decreasing both your heating and cooling costs. And of
course, caulking and weatherstripping goes without saying. You'll save both energy and dollars, and be more
comfortable winter and summer. See Chapter 6 for why
we must be concerned about saving energy to heat and
cool our houses.

Insulate hot water pipes.

Especially insulate those pipes that go along outside
walls. You could probably do your whole basement in

less than an hour, including cutting the pieces to fit and taping them together. A very simple job that even the klutziest of homeowners can do. Pipe insulation comes in various sizes, and is available at all hardware stores. When hot water has to travel all the way through the basement, upstairs, perhaps to a second or even third story, a lot of heat gets lost on the way. Lost heat equals lost energy.

Install a heat pump instead of air conditioning, or as an add-on to your furnace.

A heat pump will save you money both summer and winter. The heat pump transfers hot air from indoors to outdoors in the summertime. In the winter, it extracts heat from the outside air (yes, even if that outside air is freezing cold, there's heat to be extracted from it), and pumps it indoors at a higher temperature. A heat pump is the most energy efficient space heating system available. It could save up to 30% your present heating bills. Also check out ceiling fans as a good and inexpensive alternative to air conditioning.

If you must burn wood, install an airtight stove or fireplace insert for increased efficiency.

Forget the fireplace as a heat source. Most fireplaces operate at negative efficiency. This means that the fireplace draws in the warm air from your house and sends it up the chimney, while giving off much less heat in return. You can improve efficiency by installing a duct to provide a source of outside ventilation for the fireplace. Glass doors will also cut down on heat loss when the fire is low or out.

But while you're at it, please remember that wood is a carbon fuel, and burning wood increases the carbon in our air, contributing to the greenhouse effect. I know it's cheery to have a fire on winter evenings, but keep in mind the ecological cost. Can we, in good conscience, point the accusatory finger at Brazilians who burn the rainforests, in hope of eking out a living from the land, while we burn our own hardwood forests for the pleasure of an evening by the fire?

Avoid buying furniture or cabinetry made from tropical hardwoods such as teak, mahogany, satinwood or rosewood.

Looking for new furniture? Some beautiful woods come from the tropical rainforests of the world. Unfortunately, due to the burning of forests for cattle grazing and other questionable practices, the tropical rainforests are in danger of being destroyed beyond any hope of survival.

Many hardwood species grow in isolated groves amongst thousands of other trees. The unwanted trees are simply bulldozed aside in order to harvest the desired hardwoods. Furthermore, the tropical rainforests house millions of species of plants and animals, some so rare that they exist only in a single tree.

According to the World Wildlife Fund, three species of wildlife become extinct

**every day, primarily due to our destruction
of the earth's rainforests.**

Without our tropical rainforests, the planet really has
very little chance of surviving at all. The trees store billions
of tons of carbon, which, if released, could intensify the
greenhouse effect to the point of no return. The tropical
rainforests provide oxygen, moisture and climate control
for our whole planet. They have often been called the
"lungs of the Earth." For more on the greenhouse effect,
see the section on thermal energy in Chapter 6.

Look for locally grown hardwoods where possible.
Some examples of woods to choose instead of tropical
woods: ash, oak, beech, maple, butternut, cherry, American walnut. For more ways to help preserve the tropical
rainforests, see Chapter 15: Festivities and Gifts, and
Chapter 16: Put Your Money Where Your Mouth Is.

**Avoid chipboard, particle board, and other
glued or bonded materials which may emit
formaldehyde gas.**

Some building materials may be inherently toxic. The
resin used to bond chipboard may release formaldehyde
gas over a long period of time. Formaldehyde may irritate
mucous membranes, cause headaches and even damage
to the nervous system. And don't forget, that's the stuff
they use for embalming fluid. I wouldn't want it in my
rec room.

Urea formaldehyde foam insulation was once recommended to homeowners, but a few years later, it had to
be removed because of the health hazard. The best choice

for insulation is environment-friendly cellulose fiber made from recycled newspapers. It contains the least number of harmful chemicals, and recycles tons of newspapers, instead of using new materials.

Many twentieth-century diseases did not exist before the invention of modern building materials. Some people are profoundly affected by fumes from plastics, concrete, synthetic carpets and other materials commonly found in houses and apartments. They must live in special buildings made from and furnished with the most basic natural materials—wood, paper, stone, cotton, wool. Most of us do not react so violently to modern substances, but the fact that some people do should be a reminder to us of what kinds of materials we have chosen to live with.

Our houses and apartments may contain plastics, glues, paints, formaldehyde, solvents, preservatives, fungicides, pesticides and other toxic materials. We often do not even know what's in the new houses we buy, and we may suffer symptoms such as headaches, fatigue, or nausea, without ever linking our problems to our housing.

We may not be able to start fresh and build a new house or apartment to rid ourselves of toxic housing, but every decision we make as we change and improve our housing and decor can be more environmentally friendly than the previous decisions made.

Avoid lumber treated with wood preservatives, fungicides or insecticides.

Would you leave arsenic around your yard? Too dangerous? What about that new deck or fence? Ask your lumber yard if arsenic is an ingredient in the preservative.

Another preservative commonly used to treat wood, pentachlorphenol, or PCP, is also poison.

When I decided to build a sandbox for Caroline, the only pre-cut wood at any lumber yard was treated with preservatives. (It had the characteristic greenish tinge and etching marks where the poison was forced into the wood.) Knowing that toddlers chew on everything, I refused to buy poisoned wood for my daughter's sandbox. I ended up buying cedar lumber which I had to cut to size myself. Cedar is naturally rot- and insect-resistant. I also wanted to avoid the possibility of poisons leaching out into the soil, contaminating the yard and garden, and all the micro-organisms that keep a natural balance in the soil.

Keep wood dry to avoid insect infestation and rot. Allow adequate drainage from decking and fencing. Use varnish or wax as a preservative where possible. Choose cedar, or other resistant woods. If termites are a problem in your area, break wood-to-soil contact. Use concrete footings around fence or deck posts.

Try herbs, groundcovers, wildflowers, natural prairie grasses, native bushes and trees for a low-maintenance, environment-friendly lawn.

Experiment with natural landscaping—can you landscape your lot with no lawn at all? The attempt to grow a single species—grass—in a given area, encourages us to be wasteful of water, to try to control pest and weed invasion (often with destructive methods), and to eliminate useful species of plants and insects. We waste energy with either gasoline or electric lawnmowers, in an endless attempt at keeping the grass at an unnatural height. See Chapter 12 for more on lawns.

Ecology Park in downtown Toronto is an excellent example of landscaping without lawn. Just half a block from a bustling commercial strip, groundcovers, herbs, prairie grasses, meadows, woodlands, and wetlands play host to numerous species of life, which are being crowded out of conventional gardens in the city. This park needs no mowing and little maintenance. It features an "edible landscape" and an herbal lawn, along with wildflowers and a young deciduous forest.

Plan your renovations around existing trees.

Save your trees. They benefit our planet. Protect them from possible damage from exterior painting, sandblasting, dumpsters, or earthmovers. And, of course, when the heavy exterior construction is completed, plant trees for future generations.

Plan for energy efficiency before you start renovating.

Check with your local electrical company. You may find low interest loans available for energy efficient renovations. And remember that energy efficient improvements usually pay for themselves in energy savings— savings that can be measured in both dollars and environmental benefits.

When the walls are falling down around you, and plaster dust is in your hair, you'll feel better about the tough job of household renovations because you're following a few of these tips. That positive start will create a real contribution to your own life and the health of our planet.

10

SECOND
TIME
AROUND

Think Before You Throw

When I was a child, I sometimes made a trip with Dad to what was euphemistically called "The Nuisance Ground." This was the dump, where our town's garbage moldered and smoked with a fire that never seemed to go out, but never really seemed to burn either. The air smelled sharply of burning tin cans, acrid plastics and unidentifiable, sweetly rotting foods. Rats were said to lurk there, although I never saw any. The whole area took less land than a square block.

When I started renovating houses in Toronto, I made a few trips to the city's garbage dumps. What an eye opener. I drove my half-ton truck between endless lines of overflowing dump trucks, up a mountain of garbage, following a vague, half-formed road of mud and sludge. At the summit, garbage lay spread out like a lumpy quilt for what seemed like miles in all directions. Gulls whirled and dipped, screaming raucously, and the stench enveloped me so overwhelmingly that I whipped off my head scarf and made a face mask to filter the smell. Far in the distance, I could see green fields and trees.

Huge machines pushed and shoved the garbage in a futile attempt to organize it into orderly piles, while dump trucks continued to disgorge their loads and rattle back down the road as fast as they could go. On more than one occasion, my truck became hopelessly stuck in the muck, and I was rescued by some anonymous tow truck driver who thundered through the dump like a knight on a mission. The roar of the machinery, the screaming of the gulls, the bustle of the trucks and flying pieces of garbage all merged in a bucolic background complete with sunshine to make a scene out of some modernistic vision of hell.

We're running out of places to put garbage. For years we've complacently put our bags and cans out at the curb, secure in the knowledge that the trash would be picked up and taken away, out of sight, out of mind. Now, with our dumps rapidly filling, we're discovering that—surprise, surprise—nobody wants our garbage. Many communities have started recycling programs to try to reduce the strain on our dumps, and have closed their incinerators because of environmental and health risks. If New York and Toronto can run out of garbage dump space, so can Knoxville and Fargo, Moncton and Winnipeg. "It'll never happen here," you think. That's what everyone thought 10 years ago. Now we make nervous jokes about the garbage scow that was refused entry to every port from New York to Florida.

The U.S. recycles about 10% of its waste, and Canada recycles less than 2%, compared to as much as 51% in Japan and up to 65% in some European cities. With only 8% of the world's population, North America produces more than 50% of the world's garbage.

I'm not going to tell you to decorate ketchup jars to use for flower vases. We can all come up with ideas for "arts and crafts" that will keep the kids out of trouble for an hour or so, but have we really addressed the garbage problem? Or are we merely delaying the inevitable road to the dump? If you want to use toilet paper rolls or juice cans to make pencil holders, that's fine. But let's look at some practical everyday garbage items that we can reuse instead of junking. Especially useful are items which can substitute for things we would have bought new instead.

What To Do

Start with these ideas for uses of things you might have thrown away. Think of your own.

• Milk cartons: Barry and I cut off the tops at various heights to make square desk or drawer organizers for nails, screws, elastics, twist ties, paper clips, fuses, or the thousand and one little bits and pieces that find their way into our cluttered belongings. Although your boss may be surprised at first if you do this at the office, he or she should quickly see the benefits. Plastic desk organizers sold by office supply companies are outrageously overpriced, because it's assumed the buyer isn't using his or her own money.

• Egg cartons: Local farmers markets will usually refill your egg cartons, and will often accept used cartons for reuse. Be sure to buy your eggs in recyclable fiberboard cartons (now usually made from recycled papers), rather than in plastic foam containers. Even though CFCs are not used extensively any more in manufacturing foam cartons (making them "ozone-friendly"), the cartons are still more non-biodegradable plastic junk. I'd like to see supermarkets set up a corner to collect egg cartons for egg producers to reuse.

We also use egg cartons for sorting workshop items, storing Christmas tree ornaments and for packaging gifts. Two lids taped together make a good gift box, while the bottom half stores many small items.

• Plastic food containers: I cut the bottoms off the containers to use as cutworm guards around tomato plants in the garden. And of course, every toilet has a container full of water with a lid on it inside the tank, cutting down on the amount of water per flush. The plastic containers also make perfect storage for home made playdough for kids. Lego and other small play pieces can be neatly collected and sorted in large plastic containers.

• Plastic jugs: Some laundry products are now refillable. Procter and Gamble are promoting their "Enviropak" containers of Downy and Ivory—you use a funnel to pour the liquid into your plastic jug from the soft plastic container. Cut off the bottom half of a cleaned plastic jug and use the top half without the lid for a funnel, or with the lid for a scoop for sugar or flour. Both halves make good sandbox toys. Smooth the cut edges with sand-paper.

You can also make a tool holder by using a sharp hobby knife to cut the plastic jug into a long continuous strip, about one and a half inches wide. Tack it to your base-ment wall, or a piece of scrap lumber attached to the wall, leaving appropriate sized gaps to hang your ham-mer, pliers, screwdrivers, or paint brushes.

• Eyeglasses: Many charitable organizations send used eyeglasses to Third World countries to be sorted and distributed to needy people there.

• Plastic plant pots: Many nurseries are now accepting both used pots and flats or trays for reuse. Check with your local nursery. If they're reluctant, offer to clean the pots before returning them. Every spring I have great plans to start my own seedlings of one type or another, but usually end up buying nursery started plants. It's great to be able to get rid of all those plastic pots. I still have enough of them left in my garage that if I ever do get started on time in the spring, I'll have enough pots for the job.

• Laundry soap boxes: These sturdy boxes are often made from recycled paper, so are already one step toward being environmentally friendly. But they needn't end their useful life when they're empty. With the tops cut off, they make fabulous organizers for toys of all sizes. Cut down the sides to make storage boxes for old magazine collec-tions. Store Christmas tree ornaments in them. Cut out one large side, and use the box as a tray for starting plants.

• Clothes: Before you give up on them, see if they can be reused for another season by altering slightly, taking them in, letting them out, adding accessories, repairing or making them into something else. A shirt might become a light jacket over a T-shirt. A pair of pants may become shorts. A cozy old sweater could be reknit into a scarf.

The next step is to ask friends and family if they would like to have first choice before you get rid of the clothes. Sometimes if you're just tired of an item, it will seem brand new to your sister or friend. You may receive some treasures in return. Children's clothes are especially welcome hand-me-downs. When you're cleaning out the closets, don't forget about needy organizations such as women's or men's hostels, native organizations, the Salvation Army or Goodwill, the local church bazaar, or other charities that will ensure your clothes are used and appreciated.

Clothes too dead even to donate? Get out the scissors; this is going to be fun. Now that we're not buying paper towels in order to save our forests and rivers, we always have need for lots of rags. Shirts, nightgowns and pajamas make terrific rags. Save the buttons and zippers, and put them in your sewing basket.

• Pantyhose or nylon stockings: Both are perfect for paint strainers. When you're mixing several cans of old paint to get enough paint for a certain room, pour the paint through a layer of nylon pantyhose—it catches those lumps that might otherwise spoil your paint job.

Pantyhose cut into strips are perfect for plant ties. Unobtrusively brown, they blend in with the greenery, are strong and durable, yet soft enough not to damage the plants. I've used them with great success on tomatoes, roses and other climbing plants. If you're skilled at crafts, you can braid nylons or pantyhose in strips to make a washable rag rug for the laundry room or back porch.

• Bedsheets and pillowcases: Cut down worn sheets to make pillowcases or pillow protectors. Pillowcases be-

come great laundry bags, or good, everyday table napkins. Flannel sheets make excellent rags.

• Clothes hangers: Your dry cleaner will take them
back. They cost money, and cleaners are willing to sort
them to save the cost of buying new ones. There's no
need for any clothes hanger ever to go to the dump.

• Plastic dry cleaner bags: They're the bane of my
closets. For environmental reasons, I'm reducing the
amount of dry cleaning I do. But still we end up with
some of those wretched bags. Mom taught me the trick
of using a dry cleaner bag to roll out pastry on. It makes
the whole job much easier and more convenient than
doing it on the bare counter top. But then, she makes
more pastry (and better) than I do. I also use the bags to
cover all the out-of-season clothes in my closet.

I've seen a washable rug woven from plastic shopping
bags, although I can't say I'd prefer one over a woven
rag rug. A cushion woven from plastic dry cleaner bags
makes a waterproof seat to take along to the stadium.
You can weave a plastic square to put down on the floor
of the car to protect against wintertime slush. When we
were kids, we always used plastic bags over our socks
inside our winter boots. It made the boots slide on more
easily, and protected the socks from any leaks or stray
slush.

• Fireplace and barbecue ashes: I like to have a sack
full of ashes saved up for the wintertime. When the walks
are icy, we sprinkle them with ashes instead of salt. It's
kinder to the grass, trees and flowers, and easier on our
water systems.

• Christmas trees: Does your city or town have a wood
chipping program? Christmas trees can be chipped to
make pathways for city parks and recreation areas. A

branch of the tree makes a good boot scraper near the back door. Evergreen boughs make good mulch to protect garden plants. Tree limbs can be pounded into the compost to separate the fresh from the old. They will eventually become compost. Cut up the heaviest parts.

• Books and magazines: We've all carted around boxes of books from house to house, apartment to apartment. A day finally comes when we don't bother to unpack them any more. It's time to call the used book dealer, or donate them to the many hostels, community centers and hospitals that want them. Some overseas charitable organizations will gladly accept your textbooks.

> *Before buying, think: Do I need this? Is it overpackaged? Will it last? Thoughtful purchasing will help alleviate the garbage problem at its origins. Before discarding anything, think: Can I reuse this? Can someone else? Can it be re-cycled? How small can you make your garbage output?*

Chapter

THE
REAL DIRT

Low-tech Composting

Compost is most often described as being a "soil conditioner," but the words hardly do it justice. In fact, it's often described as "rich, dark brown, crumbly, humuslike, fluffy, earthy, sweet smelling." Compost is organic material that has decomposed into usable soil form. Simply put, it's living stuff that has become, for our purposes, earth again.

You don't really have to make compost. It pretty well makes itself. In fact, that's how nature works out in the wilds where human beings don't muck with it (in those few such places left on Earth). Leaves fall to the forest floor, plants die back, twigs and branches fall to the earth. Bacteria that are present in the soil begin to act on the fallen organic material, starting to break it down as soon as it is no longer "living." Soon fungi and protozoans get into the act, and then the insects and worms start munching away.

Think of a rotted log that has been on the forest floor for a while. The crumbly, spongy material that collapses if you step on it—that's the beginning of compost. Think of an apple that has been left to lie where it fell from the tree. It's too decomposed to eat. That, too, is the beginning of compost. In fact, anything that was once living, and is not fossilized, is a candidate for composting.

No, you don't have to make compost, but you do have to give it an opportunity to make itself. Providing the opportunity takes so little effort that I'm always astonished when people tell me they don't compost. And I'm never surprised when the newly converted tell me, with amazement, how easy it is.

We waste so much, and so unnecessarily. I get angry at the sight of good food being wasted. When I go to a restaurant or a buffet dinner or any public place that serves food, I see large quantities of perfectly good food thrown in the garbage. Even in private homes, I've seen people throw out terrific leftovers after dinner parties. Every time I see it happen, I think of the waste, and wonder if the offenders have considered either of two alternatives: the food could be refrigerated and eaten at a later time, or if they can't get over their phobia (or restaurant regulations) about "used food," they could compost the leftovers.

Good untouched food isn't the only wasted thing we find in kitchen garbage. Food that is "past its prime" or that has gone bad is often thrown out. We also throw away lots of things that we don't normally eat. Banana peels, squash seeds, apple cores, tea bags, peanut shells, carrot leaves, coffee grounds, egg shells and cauliflower leaves are just a few of the things that never end up on our table as food. Add to that list tree leaves, garden clippings, weeds and grass clippings, and our pile of garbage grows even more.

Up to 40% of our municipal garbage is made up of kitchen and garden waste.

Yet all of these things, and more, should not be garbage, but a source of raw materials for compost—an untapped gold mine. If we could reduce our municipal waste load by 40%, we would benefit immediately. We'd have 40% less garbage to haul out to the curb, and we'd also have a wonderful supply of compost for our lawns, gardens and houseplants.

Everyone in our community would benefit from reduced municipal taxes (or at least smaller increases, since undoubtedly the savings would be eaten up elsewhere). Certainly we'd be given a reprieve on the problem of our rapidly filling landfill sites.

Another important benefit of composting is the fact that anyone can do it with a minimum of effort, and without a whole mess of rules and bylaws to govern the process. It's a movement that has spread beyond the control of politicians and governments. It has put control back in the hands of the individual. You get to say what goes in your compost, what goes in your garden, what goes to your landfill site or incinerator.

And you get to reap the benefits. Unlike some of the other fruits of your labor, the results of your efforts are yours to keep. The government can't take your compost away for its own use—at least not yet. I'll tell you more about the benefits to your lawn and garden later in this chapter.

Composting is the natural process of all organic materials. But when human beings get into the act, and start burying their compostables in gigantic landfill sites along with non-biodegradable materials like plastics, the natural composting action can be delayed by dozens (maybe hundreds of thousands) of years. We lose the potential benefits of all that organic material. Your orange peels and tree leaves might as well be tin cans and plastic bags for all the good they're doing you, once they're in the landfill site.

Nor is incineration the answer. Burning garbage pollutes the atmosphere with carbon dioxide, which contributes to the greenhouse effect of global warming. If we burned only our food wastes we might have a usable ash as a by-product, but since we mix our organic waste with all other types of garbage the residual ash is full of toxic pollutants, which makes it a hazardous waste disposal problem.

If we don't compost, we're multiple losers. More waste handling, higher taxes, more agricultural or housing land wasted on garbage dumpsites, atmospheric pollution, and the loss of the organic material itself are just a few of the costs we incur when we don't compost our kitchen and yard wastes.

What To Do

Start composting today.

Don't worry about making perfect compost. Concentrate on the pleasures of the good you're doing and the benefits you're receiving. The first rule of composting is that most of the rules can be broken. Composting at the home level is a very inexact science, and I like it that way. It means I don't have to feel guilty if my compost isn't "working" perfectly this week. I know that whatever composting I do, no matter how poorly I do it, it will work at some level, and there's always an opportunity to correct the workings of the pile with the next batch of stuff I add to it.

My mom used to peel potatoes and other vegetables onto a newspaper, and then simply throw the peelings out into our big vegetable garden (I was often sent to do this chore). Eventually, Dad would till them under the earth. That was composting at its simplest level, a fine and effective method of getting rid of kitchen waste and conditioning and enriching the soil.

You don't have to wait for any special day, or make any special preparations. At our house, I just decided one day, several years ago, that I wouldn't wait any longer. I wanted to compost right then. So I did.

Your initial start up time can be as little as a minute. Start by saving your raw food clippings, leaves, peels— any stuff that looks like it was once a living plant. Put them in a leak-proof container—unless you have mice in the house, it doesn't need a lid. We keep an old plastic ice cream container in the cupboard under our sink. I wrote "Compost Only" on a piece of masking tape and stuck it to the container, just so visitors don't mistake it for a garbage can.

• **Compost everything organic**. Broccoli leaves, houseplant clippings, pear cores, tomato stems, fingernail and toenail clippings, corn husks, mango peels, cat hair, old porridge, grapefruit skins, grape stems, mushroom trimmings, human hair, cabbage cores, broom sweepings from the floors, squash seeds and housedust are just a few things that go into our compost container under the sink.

Notice any surprises in the list? I'm sure your list will be different from mine. It all depends on what you normally eat in your house. No, we don't eat cat hair or toenail clippings, but both are of organic origin, and are perfectly compostable too.

• **Chop, bruise, crush, scratch or shred materials before adding them to compost.** Chop up any large or tough pieces, such as watermelon rinds, before adding them to the pile. Although not strictly necessary, it will speed the composting process. Now that we're in the habit, we do it without thinking as we prepare our foods. When I remove a cauliflower stem, I just chop it up right there on the cutting board before I put it in the compost pail. Same thing with corn husks (I use scissors).

Sometimes, after I've finished preparing foods in the food processor, I dump any compostable clippings that I happen to have on hand into the processor (without cleaning it first), and give them a whirl. I scrape it all back into the compost pail with a rubber scraper. It really reduces the volume of your compost quickly. The idea is to give the bacteria more surface area on which to act, and to speed the composting process.

• **Choose a convenient site for your compost**. Once you have a container full, you're ready to start your outdoor compost pile. The area you need will vary depending on the size of your household and the size of your yard and garden. We're a family of three. When we had a bigger yard with more trees, we had a two-part compost pile (more on that later), with a total size of about 3' × 6'. Now that we have a smaller yard, our compost pile is less than three square feet. Larger piles will heat up and give you finished compost faster than smaller piles, but any size will work. A 3' × 3' pile that is three or four feet high is big enough to retain its own heat, which will hasten the composting process.

If you have an out of the way spot—behind a garage, at the side of the house, behind a large bush, or in a corner of your yard—you can start your compost pile simply by beginning to layer the materials. A sunny place is better than shady, but my compost pile works fine in the shade. With a smaller yard, you may want to stake off an area to keep the compost more neatly contained and clearly defined. For apartments or very tiny yards, see the sections on vermiculture and garbage can composting, below.

Depending on how much effort you want to put into it, you can "wall" your compost with snowfencing, cement blocks, or screening. Your enclosure may be square or round. But it's not essential to make walls for your compost. It will work just fine as a pile without walls. If you do decide to build an enclosure, allow for air circulation. Snowfencing or chicken wire will hold the compost in while allowing the air to circulate. If you use cement blocks, stack them with the holes going sideways instead of vertically, or stagger the blocks, leaving spaces between. You may wish to make a door, or eliminate the fourth side of the enclosure, to allow easy access to your pile.

If your soil is very sandy and loose, you can dig a shallow pit for your compost, simply to give it a lower profile in your yard. But beware the too-deep hole. Four to six inches is plenty. Any deeper than that, and you may have problems with drainage and air circulation, which will make your compost too wet. Put some larger sticks at the bottom of the pile to encourage circulation. At least part of the pile must be above ground level.

When we had our bigger compost pile, we dug a shallow pit. Then I wedged some scrap strips of plywood about six inches higher than ground level all around the inner edges of the pit, and held them upright with small stakes. That gave me some height to work with, and defined the edges of the compost. We had very sandy soil, so drainage was never a problem.

• **As you collect materials, just start piling them onto the compost pile**. If you have a supply of leaves to start with, so much the better. Leaves are excellent for composting, and turn into that beautiful dark brown leaf mold that smells like a forest. Any kind of leaves are fine, even evergreen needles or oak leaves, which are very acidic. I've used bushels of oak leaves without harming my compost. Leaf mold is an excellent conditioner for clay or sandy soils, because it can hold 300% of its weight in water.

• **Add as much organic material as possible to get the pile "cooking."** After your layer of leaves, throw on your kitchen scraps and any other soft clippings, such as grass or plant leaves from the garden. If you have some dirt available in your garden, throw on a layer of dirt. This provides the local micro-organisms that begin multiplying and breaking down the material immediately. Try to build up your pile with organic material. If you don't have the material available, just cover with a thin layer of dirt and wait until more becomes available. Sprinkle the pile with water, but don't soak it. Turn the pile over with a shovel or pitchfork on the second or third day, and once a week after that. Try to alternate the layers of materials, so that any one material doesn't become predominant. This will accelerate the breakdown process.

• **When you have enough material, the pile will begin to "cook"—that is, it will heat up to the temperature that will begin the composting process**. The cooking will destroy weed seeds and disease organisms that may be present, and break down the organic material.

• **Mix roughly equal volumes of carbonaceous and nitrogenous materials.** A good guide for the mix of materials is what's called the "C to N ratio." That means the balance between carbon and nitrogen in any given organic material. A rough guide is that things which are carbonaceous feel more dry and woody, like dry tree leaves, while nitrogenous material is softer, mushier or easily bruised, like grass clippings or most kitchen wastes. Some scientists suggest very complex balances of certain types of materials, but again I stress—don't worry if yours isn't exact. It will still work.

Approximately equal *volumes* of the two types makes a good compost. For example, mix one pail full of kitchen wastes with one pail full of dry leaves. The nitrogen will ensure quick composting, while the carbon will balance the mixture to prevent it from rotting and giving off that stinky ammonia smell. You will always have a greater weight of nitrogenous material because it is denser. A pail full of grass weighs a lot more than a pail full of twigs.

But all this C to N discussion is more precise than the way I manage my compost. I'm just not that fussy. I'm sure I don't spend more than five minutes a week, if that, either thinking about my compost, or working on it. Believe me, you won't have to either.

• **Expect your pile to shrink quickly.** Sometimes Barry confronts me with an accusatory tone, and demands, "Have you been using that compost?" We always laugh, because it's there for us to use whenever we're doing any gardening. But the funny thing is that compost shrinks terrifically, even when you're not stealing from the pile. A three-foot-high pile can shrink to half that size in a few weeks, just from the decomposition that occurs. So don't be discouraged if your pile is enormous and you seem to have too much of the stuff. In a very short time, you won't have enough.

You should be able to feel the heat of your compost working if you dig your hand into the center of it. If the pile is not cooking, it's easy to correct by adding the proper ingredients. If it stinks and is wet, add some shredded leaves or a pail full of dry soil, or even some shredded paper. If it's very soggy from heavy rainfall, or over-enthusiastic watering, do the same. Also turn it every day to help it dry out.

• **Correct the mixture if it's not working**. If your compost has no smell and it's still not cooking, it has too much carbon material, such as leaves or twigs or paper. You need to add kitchen wastes or grass clippings to give it more nitrogen. If it's very dry, add enough water to dampen it.

If flies, fruit flies, wasps or other insects are attracted to your compost, it's likely because you haven't covered the kitchen scraps with dirt or leaves. In all the years I've been composting, I've never seen flies anywhere near it. Living in the middle of the city, I've always had the compost right beside the fence, with no complaints from any neighbors. Animals will not be a problem if you avoid putting meat scraps or bones in your compost, or bury them deeply inside the pile. If you want to put in a few bones, they will eventually decompose, but you'll likely keep running into them for years as you dig out your compost. You can just keep throwing them back in if you wish. Any larger materials such as twigs that haven't broken down completely can be screened out of the compost, or simply picked out by hand and thrown back into the pile.

• **Divide the compost into old and fresh piles**. Every time you add new material, you have to wait for it to cook until you can use the finished product. So if you're anxious to get using your new compost, you may wish to start a two-part pile with a plywood or screen divider. One half will be compost that's ready to use, the other half will be the pile to which you add your fresh materials. You can keep alternating which side you take from, or you can throw some of the newly finished stuff onto the other side before you add fresh materials.

• **Make friends with worms**. If you think you don't have earthworms in your yard, just make a compost pile. It's amazing how they find it, and do they ever get fat and multiply. Barry says our worms are starting to look like small snakes. Earthworms are excellent composters, chomping their way through tons of organic material, turning it all into dark, rich castings.

Although I'm not the squeamish type, I never felt comfortable handling worms. Over the years I've been composting, I've grown to accept and even like worms. Although I still don't relish it, I can now pick up worms, even with my bare hands, and recognize them for the friends they are. If you feel that you aren't ready to make friends with worms, wear garden gloves when you handle your compost, because you're sure to run into a few worms from time to time. You might even come to like them. If you have any fishing friends, they'll be delighted to take a few worms off your hands.

When is compost ready? It will take anywhere from a couple of weeks to a year, depending on the volume and type of materials you add to the pile. When you think it's ready, scoop up a double handful.

You'll know it's done when it looks like dark brown loam, is crumbly, moist, loose and fluffy. It will smell wonderfully earthy. You won't see any sign of the original materials. You've got compost.

• **Expect some bonuses from compost**. Sometimes the original materials can pop up in unexpected ways. Once we dug in several pails full of compost around some hedge bushes in the spring. Later in the summer, I noticed what seemed to be a squash plant growing up the hedge. We decided to leave it and see what we got. We ended up with six or eight delicious squashes grown from seeds which hadn't cooked in our compost. Like I said, it's an inexact science. I've also had volunteer tomato plants spring up, but without ever producing fruit. Some people actually plant tomatoes or other vegetables around the edges of their compost pile, where they won't get too hot. If you have a big enough compost pile, it makes a terrific growing medium, and the plants help to hide it.

• **Save a pile of leaves in the fall to mix with your kitchen scraps in the spring**. Remember that C to N ratio? You'll need something dryer, more carbonaceous, to go with all the soft material you save over the winter. You can save a plastic garbage bag or pail full of fall leaves—they will have started their own decomposing inside the bag by spring time or you can simply make a pile of leaves in an out of the way corner.

The warmer the weather, the better your compost pile

will work. It doesn't matter where you live, you can continue your composting efforts all winter long. It freezes in Toronto, but that doesn't slow me down. In fact, it's even less work—if that's possible—in the winter time. The actual composting process itself will stop entirely in freezing weather, but as soon as spring comes, you'll be ready with a nice pile of stuff to start your pile cooking again.

• **Don't stop collecting compostables in cold weather.** We continue to collect our kitchen scraps all winter long in the container under the sink. We put a plastic garbage can outside our kitchen door. (You can put it on your veranda, in your garage, or near your compost pile, but in the winter it's nice to have it handy so you don't have to bundle up just to put out the compost.) We dump our small container into the bigger one outdoors, and just let it freeze.

I usually leave the lid open a crack, so the stuff doesn't start to rot and stink before it freezes. You should also throw in a little dirt if it's available. If raccoons or other animals are interested in your can, you might have to put a weight on top of the lid while the can is relatively empty to prevent them from knocking it over and scavenging. When it starts to fill up and freeze solid, nothing can budge the container.

It always amazes me how, even in winter, the pile of stuff shrinks and settles, so that one garbage can will last our family through all the freezing months. If you have a large family, you may need a large-size garbage can to save your winter kitchen wastes.

Don't worry if you get a few warm days in the winter. Your solid lump of waste is unlikely to thaw. When spring comes, your pail will stay frozen until you can't wait any longer to get out into the garden. This is likely to be the only time when you have to endure a bit of stink. Your pail of wastes has started to decompose anaerobically, that is, without air, and that's what makes it give off that ammonia odor—but not for long.

• **Layer your winter kitchen wastes with the leaves you saved from last fall**. Add a layer of dirt if possible. Any odor from your winter waste can will disappear very quickly. You may need to turn the pile frequently to dry it out the first few days. If you have some compost left from last year, you can throw on a few shovelfuls, and that will also help to quell any odors.

What Else Can Go In The Compost?

• Fish bones, being less dense than meat or poultry bones, will disappear more quickly in your compost. As with other bones, dig them well into the center to avoid animals rooting in the compost. A little fish skin attached is fine too.

• Empty your vacuum cleaner contents directly into your compost pile. Almost everything in your vacuum cleaner is of organic origin. Believe it or not, much of household lint consists of dead cells of human skin tissue. Some of the other stuff in your vacuum is cotton or woolen lint, hair, food crumbs and dust. All these materials are totally compostable. I find it a great convenience to be able to dump from the vacuum cleaner right onto the compost pile. If you have a vacuum cleaner that uses disposable bags, you can compost the paper bag at the same time, although you should tear it open and rip up the paper. Depending on the design of your bag, you may be able to reuse it instead of disposing of it. Just shake the contents into the compost.

• Hardwood ashes from fireplace or stove are an excellent "sweetener" for the compost and help the pile work faster. They're high in potash, which will help to make your plants more vigorous. If you don't have any ashes available, you can ask your friends to save you a box full the next time they clean out the fireplace. Failing that, you can buy a box of dolomitic limestone at your garden store or hardware store, and sprinkle a few handfuls on every now and then. Don't use lime at the same time as manure. Their interaction will cause loss of nitrogen.

• I've already mentioned hair, but it's worth repeating. Whenever I clean out my hairbrush, I just throw the hair in the compost container. Any clippings from home haircuts go in the pail. Our cat Goldwyn has very long black hair, and tends to shed quite a bit. I'm always scooping up a handful of cat hair from his bed. That too goes into the compost. In fact, any pet hair is compostable—dog, rabbit, gerbil, whatever. Bird feathers, provided your bird is healthy, also qualify.

• Sawdust is highly carbonaceous, so must have a great deal of nitrogenous material mixed with it to compost readily. On its own, it will break down quite slowly. It's a good additive if your pile is stinky or too wet, but if you generate large quantities of sawdust, you may want to start a separate pile of sawdust to work over a few years, rather than weeks or months. Be sure to keep it damp, and if possible, add some nitrogenous materials to it to hasten decomposition.

• Weeds which are not in seed are fine in the compost.

They will decompose just like any other plant material. If they are in seed, you can still put them in, but only if you're certain your compost is reaching a hot enough temperature to kill them. Otherwise, you could spread weed seeds when you spread compost in your garden. If you won't be using your compost for several months, any weeds that may spring up in the pile can be pulled and dug back in.

• Pine cones, maple wings, twigs and other woody materials are carbonaceous, and may be used in addition to leaves. You may want to break up twigs. I find that if I use twigs larger than a pencil in diameter, I'm always sifting them out, because they don't compost as rapidly as the rest of the material.

• Organic fertilizer may be added in thin layers to hasten composting. Chemical fertilizers may not suit your earthworms, and may be harmful to the natural organisms working in your compost pile. Do not use "weed and feed" fertilizer.

• Well-rotted manure is a good source of organic matter, and will add nitrogen, phosphorous, potash and other important elements to your compost. Although lower in nutrients than fertilizers, manure is valued for its soil building and conditioning qualities. If you can't get it from a farm, it is widely available at garden stores, and even supermarkets, in dehydrated and pasteurized form, for a relatively low cost. Add a shovelful to your compost pile every now and then.

• Pet manure is perfectly safe to use on compost which will not be used for food gardens. Some experts suggest letting it compost for a full year before using it. Since it might attract other dogs, you may not want to use it on an open compost pile. Opinion is divided on cat litter. The litter itself is compostable, but there's a possibility of transmitting some disease organisms from cat manure, so again, I suggest using it only on ornamentals, rather than on food crops. Rabbit manure and hen manure are high in nitrogen, and must compost for a month or so before use.

• Charcoal ashes from barbecues are compostable, although some experts advise against using briquet ashes, because of the possibility of unknown compounds being present. I usually use the old fashioned "lump" charcoal, which is just made from chunks of wood, and is a good conditioner for the compost.

• Wool, cotton, linen and silk scraps will eventually compost, all of them having come from organic sources. No polyesters or nylons, however. Cut the scraps up into very small pieces to hasten their decomposition. You may run into pieces of fabric scraps regularly as you work your compost if you use a lot of them. I once put some short pieces of wool yarn into the compost, and they were still recognizable a year later. Eventually they disappeared.

• Garden clippings of all kinds are excellent compostables. Everything from bean stalks to tomato vines to rose petals can go directly into the compost. Woody material, such as hedge-clippings, should be cut into small pieces and mixed well with wet compost. If you have too much of it, you may wish to compost it separately. Take care with garden clippings that may be diseased or infested. Botrytis blight, smut, black rot, black spot and borers can all be transmitted through compost unless the pile has stood for a couple of seasons. If you use your pile more frequently, you may wish to burn diseased clippings, or otherwise dispose of them. If your barbecue is still hot some night after supper, you can put a small quantity of diseased clippings in to make use of the leftover heat. Compost the ashes.

• Onion skins, garlic skins, grapefruit peels and orange peels are not an earthworm's dinner of choice. However, if used in moderate amounts, and well mixed with other materials, they'll work fine in your compost. I've never had any problems with the normal amounts that we put in ours.

• Old potting soil from houseplants may contain vermiculite or perlite or any number of additives. All will be perfectly safe for your compost pile. Sometimes when you don't have any dirt available to cover the latest addition of kitchen scraps, an old houseplant pot full of dirt comes in handy. Of course, the dead or dying houseplant is compostable too.

Helping The Compost Along

• Help compost to cook rapidly. One of the handiest aids to composting is any tool that will help cut up or shred the raw materials for your compost. Try a blender or food processor in your kitchen—or simply get in the habit of chopping up your compostables as your create them in your kitchen work.

• Reducing the size of yard and garden waste is a bigger chore. You can purchase a leaf shredder for about $170 in home centers or hardware stores. It has a hopper into which you throw leaves and small sticks. If you have a small quantity of autumn leaves, you could try this trick:

Fill a large garbage can with dry leaves. Hold an electric "weed eater" or lawn edger (the type that has the flailing plastic string) inside the can, and turn it on. Works like an oversized blender. I've tried that method, but found it slow compared to my electric lawn mower.

We simply pile the autumn leaves on a tarp or old sheet, and drag them as close as possible to where we want to compost them. We run the lawnmower over the pile of dry leaves until they're reduced to very fine confetti-like pieces. After mowing, we're left with a terrific material that's handy to sprinkle over fresh additions to the compost pile, winter or summer. Be sure to wear a mask when shredding dry leaves by any method. The dust really flies.

I find composting the leaves a lot less work than bagging them in garbage bags. Our city is now asking residents to put autumn leaves in clear plastic bags so that the city can identify them at the curbside and take them for composting. That's a good start, but I still find it less work to compost my own.

• Compost "starters" or activators found at garden centers contain various enzymes and other organisms. If you have absolutely no soil available to get your pile started, you may want to buy a starter, but it shouldn't be necessary if you can sprinkle on even a little bit of dirt. Your local dirt will contain all the organisms necessary to get your pile cooking. The more organic material you feed them, the more the organisms will multiply.

• Garbage cans, plastic drums or plastic bags make handy enclosed compost containers.

Drill holes in the garbage can or drum, or poke some in the plastic bag to allow for air circulation. Place in a sunny spot if possible. The compost may be wetter than a compost pile would be, and may develop odor if the can or bag is airtight.

• Plastic compost bins are expensive—anywhere from $80 to $200—but great for people who have a very small yard, or no place to put a compost pile. Some municipalities have provided bins to homeowners for free or for a very small fee. The plastic bins keep the compost tidy, enclosed and protected from animals. Some people also think they look nicer.

Vermicomposting

• If you don't have a yard or garden, but have a little space available, say a corner of an apartment balcony or a small table in your office or classroom, you can compost too. Practice vermicomposting—composting with worms. All you need is a bin with a lid, some bedding such as grass, straw, leaves, or even shredded paper, and some earthworms—preferably Redworms or "red wigglers."

• The bin can be a plastic lidded box such as the type sold by moving companies, a foam picnic cooler, or even a sturdy cardboard box with a plastic liner. Drill about 10 small holes in the bottom of the box for drainage. Set the bin on some blocks or bricks with a tray underneath. An old boot tray or a cookie pan would be fine.

• Fill the bin with organic bedding material and keep it as damp as a wrung out sponge. Redworms will eat their own weight in bedding every day. A bin of about three cubic feet can take up to a pound of worms. Order them from a worm supplier—check your Yellow Pages under "fishing bait," or borrow some from a friend who has a garden, especially if your friend has a compost pile. You'll be able to pay back your friend in worms within a year, or give some to another friend who's starting vermiculture.

• You can feed your worms the same kinds of food scraps as mentioned above in regular composting, but avoid large quantities of meats and fats. Just dig the scraps into the bedding material.

• Don't worry about the worms "escaping" into your apartment or office. They have no interest in getting outside their cozy environment. In fact if they are taken outside their box, they will die from lack of food and moisture.

• Protect the worms from extreme temperatures. If the temperature drops below 5°C or 40°F, take your bin indoors for the winter. And if it gets above 27°C or 80°F you may want to take the bin indoors for fear of cooking your worms to death. Indoors, the bin can go under a table, in a closet, or on a shelf, or anywhere you might put a cat litter box.

• Harvest your worm compost. When you no longer recognize the original bedding material, push the compost to one side of the bin and fill the other side with new bedding and food wastes. The worms will move toward the new source of food. After a week or so, gently dig out the finished compost from the first side. Dig carefully, one layer at a time, to avoid digging out your worms.

• Be prepared for troops of visitors wanting to see the worms, discover how it all works and find out how they can get started in vermiculture.

What To Do With Compost

• Dig the compost into your garden. You can never have too much of it. It's a wonderful soil conditioner and organic enricher. It will break down clay soils, improve water holding capacity of sandy soils, and add nutrients to all soils. Your plants will show an immediate improvement if you haven't been using compost before.

• Screen it and sprinkle it on your lawn. You can use compost to fill hollow spots in the lawn, or just to top dress. Don't worry if it it looks like a dirty lawn. By the time of the next mowing, you won't see it any more. If you're anxious to make it disappear, you can water it in.

• Mix it half and half with potting soil for houseplants. Most houseplants will benefit from soil conditioning, especially if they haven't been repotted for a few years.

• Try compost as a superb mulch on garden plants. Use it straight or mix it half and half with manure. Mulch helps to conserve moisture, reduces soil erosion, controls weeds and extends the growing season. But beware of crown rot in perennials during periods of heavy rain. The mulch can hold so much moisture that the perennials will rot at the ground level. Avoid the problem by leaving a bare ring around the base of the plant.

• Bag excess compost. If you are such an avid composter that you end up with more compost than you can use (a highly unlikely scenario), you can bag it as gifts to gardening friends, sell it at your yard sale, offer it to your neighbor, or put it on your front lawn with a sign reading "Free Compost." It will be gone within the day. Guaranteed.

Is all this more than you ever wanted to know about Compost? If it is, here's this chapter encapsulated: "Mix organic materials in a pile with some dirt. Keep moist. Turn occasionally."

And that's it. Remember the first rule of composting: most of the rules can be broken. It can be as simple or as complicated as you wish. Don't be intimidated, just dig right in.

> *If everybody practiced composting their kitchen and garden waste, we could reduce our worldwide garbage problem by billions of tons per year.*

12

KINDER, GENTLER GARDENING

Listening To The Earth

Synthetic fertilizers, herbicides, pesticides, fungicides and insecticides have, over the past few decades, become staples in the gardener's shed. Got a bug? Zap it with a poison. Plants too small? Hit them with fertilizer. No problem too big or too small to control with chemical help. Unfortunately, our indiscriminate use of poisons and other chemicals is beginning to catch up with us.

> **According to Greenpeace, we use four billion pounds of pesticide on our planet every year.**

Have we forgotten the lesson of DDT? Years after it was banned in North America, traces of the deadly toxin continued to show up in animals as far away as Antarctica, and in human breast milk in every country of the world. Why was it allowed in the first place? The answer is we thought it would do more good than harm. We didn't consider the long-term effects of contaminated water, air and soil. We didn't ask what the effects would be on micro-organisms, insects, animals, birds and fish in the food chain. We didn't think before we sprayed.

I wonder how many garden and agricultural chemicals that we use today will become the DDTs of tomorrow. We seem to be ignoring the lessons already painfully learned.

The problems with poisons sprayed indiscriminately on plants, soil and in the air are numerous. Humans exposed to pesticides may sustain damage to liver, respiratory systems and kidneys. Birth defects, cancers and neurological disorders have also been linked to pesticide exposure. Exposure may trigger attacks of other conditions such as asthma.

Pesticides may leach from treated lawns, gardens and farmlands into lakes, rivers and underground springs, the sources of our drinking water. From there, it's only a short step to pollution of the oceans. Pesticide and fertilizer run-off are already cited as major sources of contamination on North American coasts, the Caribbean, Southeast Asia, the Japanese coast, the Baltic, the Mediterranean and other seas.

The manufacturing of chemical pesticides may itself be hazardous. Certainly there are dangers wherever large quantities are stored. Witness the disastrous leakage of toxic gas from a pesticide plant in Bhopal, India. Twenty-five hundred people died, and thousands more suffered serious injury.

Used pesticide containers contaminate landfill sites, and are often left exposed to the weather, their residue gradually contaminating the environment. Last summer I saw dozens of empty pesticide drums sitting uncovered and unattended beside a drainage ditch, by the edge of a wheatfield on the Canadian prairies.

Animals, birds and fish eat insects and food crops, and drink water, all contaminated with pesticides. If they do not die, these animals may be consumed higher up in the food chain, concentrating the pesticides at ever more dangerous levels. Birds' eggs may have shells too thin to survive. Fish may become unsafe to eat. Mothers' milk may become dangerous to nursing infants.

It's not only agricultural pesticides that are poisoning our land and water. We're doing it in our own back yards.

Wildlife toxicologists say that songbird poisoning from lawn-care products is a little-studied environmental problem that has only begun to surface this decade.

But these are not the only pitfalls of crop and garden chemicals. Some hybrid species have lost their natural resistance to disease and insects, and require constant spraying for survival. Many insects have become resistant to pesticides, causing even greater crop damage.

Most insects are either beneficial or harmless in home gardens, yet most pesticides kill indiscriminately. Remember that even termites are absolutely necessary in the great scheme of things to keep the Earth renewing itself. The beneficial insect that would have eaten the unwanted insect is wiped out, creating further need for pesticides. Nitrogen-fixing bacteria in the soil may also be affected, further depleting the soil.

All successful yards, gardens, parks and farms contain a wide variety of plant and animal life, including insects, weeds, and unseen bacteria and other micro-organisms. When we upset the complexity and diversity of a given area by attempting to make it support only a single organism (such as grass for a lawn), we are attempting to trick nature into an unnatural act. It never works, even though we keep trying.

We can attempt to get rid of slugs, aphids, dandelions and crabgrass. But rather than blanket killing of all insects and all weeds, we should consider ways of attacking the problems individually. Otherwise, we could end up killing ourselves.

> **In developing countries, 500,000 people are poisoned and up to 20,000 die annually from using agricultural chemicals. A study of farmers reported that 1 in 10 show symptoms of poisoning.**

Around the world, we are using chemical fertilizers and pesticides in an attempt to make depleted soils produce crops that will further deplete the soil. The land and surrounding animal and plant life are greatly affected, and the results are less than satisfying. But there is good news. Some countries, such as China, emphasize organic farming, biological controls and soil conservation, which enable them to feed their enormous population successfully.

Home gardeners and organic farmers in North America, Britain and Europe are beginning to make an impact on the way we eat, and what we do to our soil, air and water. In many cities we're even beginning to see major supermarket chains advertising organically grown produce—stuff that a few years ago we could only find in small natural food stores and co-ops. It's not a fringe movement anymore. It's big business.

Whether you grow a huge vegetable garden or just a few tomatoes, whether you have acres of lawns and dozens of flower beds or just a few petunias, you can ensure that you make a meaningful contribution to the improvement of the air we breathe, the food we eat and the water we drink.

What To Do

Set your mower blades high.

Don't scalp your lawn. The absolute minimum height for grass is about two and a half inches. Three inches is even better. The short stubble that some homeowners seem to prefer is harder to maintain, encourages weeds and disease, and requires more intervention.

Take the grass catcher off the lawn mower.

The earth, rain and sun will reward you with free compost that you don't have to sprinkle on your lawn. If your grass clippings turn brown and look like hay on the lawn, you haven't set your mower high enough. If you let the grass shoot out the mower on the side that hasn't yet been cut, it will get chopped up even more with each pass across the lawn. Your grass clippings will be less than two inches long, and they'll fall through the lawn to the earth, where they will decompose rapidly. And it's less work.

If you've really let it go too long, and decide you must collect the grass clippings, be sure to put them on the compost pile. When you have a great heap of grass, it's better to mix it with some looser material like dried leaves or fine twigs. Turn the heap frequently to let it dry out. This will prevent the grass from stinking as it decomposes.

Give up some lawn for a mixed garden— flowers, trees, ground cover, vegetables.

Think about how much lawn you really need. A lawn is an unnatural ecosystem. By trying to maintain a single species—grass—over a large area, we're fighting nature's love of diversity. Just as a virgin forest thrives and supports thousands of species of plant and animal life, nature constantly strives to defeat us by introducing weeds, insects and other plant and animal life to our lawns. If we kill everything but the grass, we lose the great variety of life that is necessary to keep our planet healthy. The bonus for reducing your lawn: less mowing, less maintenance, less watering, and more interesting and varied scenery out your windows.

Resist the temptation to smother the earth on your property.

Lawn is preferable to the paving stones, brick, asphalt, or concrete that some people put in front of their houses. Every bit of greenery on our planet helps cleans the air, contributes to our oxygen supply, helps prevent soil erosion and makes our lives more pleasant.

Spend an evening outdoors weeding your front lawn by hand.

If you do it with a friend, you'll have the bonus of an evening of conversation. The first warm evening, after the weeds have reared their heads for the season, take a friend outside to dig weeds with you. Next week you can do your friend's lawn.

It's also a guaranteed way to meet your neighbors and get lots of free advice. When we moved to an old house years ago, we inherited a postage stamp city lawn full of plantain weeds. Barry and I went out one evening armed with short handled weed diggers, and just sat on the front lawn and chatted while we worked. We cleared up all the plantain in a couple of hours, met every neighbor on the street, and began some friendly relationships that have lasted for years. At another house with a bigger lawn, we did the same thing, and met another bunch of neighbors. The interesting thing is that each time we go out and hand weed, somebody comes along and thanks us for it. They don't like to see people spraying chemicals in their neighborhood. I've never yet seen anybody thank a neighbor for spraying pesticides or herbicides on their lawn.

Spend 20 minutes a week digging weeds early in the season, and you'll reap the rewards later in the year. If you have a small city lot, you can go around the yard with a plastic bag, slip it over a dandelion to prevent the seeds from blowing away, and then pull up the dandelion. The kids might enjoy this game. You should still dig out the roots. You don't have to spray chemicals.

Hand pick insects early in the season.

Hand picking is an excellent way to control insects at all stages of development. If done early, and often, you can usually eradicate the worst offenders before they start to reproduce too vigorously. If you feel squeamish about squishing bugs or slugs, use garden gloves, take a twig in hand, and knock the pests into a can of water with a little detergent in it. An early morning stroll through your garden with a friend is another opportunity for a visit while you go about your pest control. Reward your friend's assistance with a pest-picking visit to his or her garden.

Remember that most insects are beneficial.

Spraying poisons harms birds, earthworms, soil, air and water. Some insects are especially friendly and useful around the garden. One ladybird beetle, also known as the ladybug, may eat up to 1000 aphids in her lifetime, not to mention mealybugs, plant lice and mites. Dragonflies eat hundreds of mosquitoes each day.

Spiders catch numerous flying and crawling insects. When I was about eight, my Aunt Lenore taught me to catch spiders indoors on a piece of paper, and set them down outside. To this day I never kill spiders, indoors or out. Sometimes if I find them in the house, I put them in the basement, and sometimes outside. If you can't learn to love them, at least learn not to hate them. They're totally beneficial. I have no idea why Miss Muffet was frightened away.

Put a cone of birdseed in your garden.

Birds are a lot more efficient at killing bugs than we are. Purple martins, well known for their voracious appetites for mosquitoes, nowadays depend on human beings to build their homes. A 120-foot-high martin tower in Louisiana with 1116 nesting cavities is the largest martin condominium in the world—but we needn't go to those lengths.

If you have the space, one or two purple martin homes make a wonderful addition to the back yard or cottage. Check your library for woodworking plans to build your own, or see Chapter 15's Green Gifts for where to send for plans. A hollowed gourd hung high in a tree will work too.

Flickers, warblers, finches, jays, robins, grackles, sparrows, cedar waxwings, starlings and numerous other birds will polish off literally thousands of insects every day. One warbler can eat 6000 aphids in one day. A flicker can eat 3000 to 5000 ants per day. If we poison the insects, we run the risk of poisoning the birds that control them.

Plant flowers or a shrub that will attract birds.

Evergreen cones provide seeds, while the trees themselves provide nesting places for birds. Any shrubs or bushes noted for seeds or berries will attract birds. Try pincherry, white flowering dogwood, honeysuckle, holly, white pine, Russian olive, or any shrubs that your local nursery suggests for the purpose. Don't worry about the birds only eating the berries and ignoring the insects. They'll come for the berries, and stay for the bugs. Sunflowers and marigolds will also bring feathered friends to your garden. The added bonus: you have the pleasure of watching birds, and knowing you're not spreading poison around.

Try organic pesticides once.

Ask yourself before you spray: what good are all the perfect roses in the world if we're too poisoned to enjoy them? What's the use of having a perfect lawn if you can't let the kids on it because of the chemicals?

Here are some simple home recipes that you can whip up in your blender. The first one will knock the stuffing out of aphids, white flies, woolly aphids and red spider mites, and won't poison our planet. The main ingredient? Garlic. I'd try it on any insect infestation, whether in the garden or on houseplants, before trying anything else.

Chop two or three bulbs of garlic very finely, or use a garlic press or food processor. Not cloves, but the entire bulb. Pour on a tablespoon of vegetable or mineral oil, mix it all together, and let it sit for a day. In a separate jar, dissolve 1/4 cup of soap flakes or a few slivers of hand soap in two cups of boiling water. This is a good way to use up those ends of soap. The next day mix the garlic and soap mixtures together. Strain the liquid (I use old nylon pantyhose, but you can use cheesecloth or a kitchen strainer). Store the liquid in a tightly sealed and well-labeled glass jar. When you're ready to spray, pour half this pungent mixture into a gallon of water. Safe to spray on all plants, I have used it successfully on hanging pots of impatiens.

There are many variations on this spray. Try just chopping up raw onions or garlic into a purée, and soaking it in water. Add a little hot pepper sauce for extra kick. Or chop up hot jalapeño peppers in water. To save effort, do it some day when you've got your blender or food processor out anyway. Spray the liquid directly on plant leaves.

A sticky spray for spider mites is a mixture of two cups of flour and ½ cup of buttermilk mixed in two gallons of water. Spray both sides of the leaves and repeat in ten days.

Soap and water will kill many "soft" insects, such as aphids. Try ½ cup of soap power (such as Ivory Snow) dissolved in eight cups of hot water, or some old slivers of hand soap soaked in boiling water. A safe and effective spray.

• Slugs and snails: Home remedies abound. The most popular is a saucer full of beer set in a hollow so the rim is level with the ground. Slugs will crawl in and die a happy death. If you're doing regular pest patrols of the garden, carry a salt shaker. Simply salt the slugs as you see them, and watch them explode. Instant death. Not for the faint of heart. If you want a less random way of spotting slugs, leave a flat board or grapefruit or orange skins, cut side down, overnight on the garden. Lift in the morning and salt the slugs underneath. A preventive— sprinkle wood ashes or barbecue ashes around the base of plants. The slugs won't drag their slimy little bodies through the dry ash. Must be repeated after every rainfall or watering.

• Chinch bugs: The nymphs of these bugs suck grass blades, making large patches of lawn turn yellow, then brown. Mix a pail of water with a cup of laundry soap flakes like Ivory, or ¼ cup mild dishwashing soap. Pour the mixture over the affected area. Put an old sheet or towel over the area and watch the bugs jump on the cloth to get away. Then you can just dunk the cloth in a bucket of the soap mixture.

• Earwigs: May be eating larvae of harmful insects, so take care in your attack. Make sure they're a problem before you decide to eliminate them. Attract earwigs with old rags left on the ground overnight. Simply shake out the rags and step on the bugs you find. Lay an old cardboard mailing tube, a rolled up newspaper, or a piece of bamboo on the ground overnight. Dump earwigs onto the sidewalk and squish in the morning—or dump into a pail of soapy or detergent water (yes, you can even use your dishwater from breakfast).

• Wasps: Again, may be doing more good than harm. Some species of tiny wasps are voracious predators of harmful insects. Only if the wasps are a terrific nuisance when you're trying to eat or work outdoors should you try to discourage them. If you have a nonaerosol hairspray, a quick blast will put wasps out of commission. If you can spot their nest, wait until night, when all or most are inside, spray shaving cream on their entry hole, and slip a plastic bag over the entire nest. Close the top tightly, and dispose of the entire package.

An old but effective remedy, if you have a big enough garden, is to hang a piece of meat or fish skin far away from the area in which you're working or eating. Place a pail of water under the bait. The meat will attract the wasps away from you, they will gorge themselves on the meat, fall into the water and drown. Sounds bizarre, but don't laugh until you've tried it.

• Caterpillars: One of the most ingenious controls I've

heard about involves vacuuming caterpillars from your trees. I haven't tried it, so only pass it on as a spur to inventive gardeners. One resourceful thinker used duct tape to fasten a 10-foot length of PVC pipe (plastic plumbing pipe) to the end of her vacuum cleaner, so she could stand on the ground and vacuum her tree. Vacuuming could work equally well with other insects. Dad used to vacuum flies off the windows, and plug the end of the hose with a wad of paper to prevent them finding their way out again. Be sure the insects are actually dead before dumping the vacuum contents into the compost pile.

• Flies, fruit flies, moths: Check your garden center for various traps, funnels and tents, all meant to capture flying insects. Some of the traps release pheromones (sex attractants), which disrupt the mating cycles of insects. Others work by capturing the insect in a trap where they will drown in a detergent/water solution, or stick to a coated strip. Beware of sticky strips coated with pesticides. That's literally overkill.

• Cutworms: Put collars around tender young transplants or seedlings. I always use old plastic yogurt containers, cut into rings about four inches high. Push the ring into the soil about two inches deep. Tin cans with tops and bottoms cut out also work well. This preventative measure provides effective control against cutworms gnawing off young shoots before the plants mature.

• Ants: Sprinkle their hills with borax—available in the laundry detergent section at your supermarket. Or try diatomaceous earth, available from your garden center, but watch out for ones containing chemical pesticides. Mom pours boiling water on ant hills, with some success. Ants won't really harm much of anything, except possibly your lawn. They are completely harmless to your peonies.

You'll see them crawling on the peony buds before they open, attracted by the sweet nectar, but it's no cause for alarm. If you wish to take peonies into the house for cut flowers, just wash them down with a light spray from the hose.

Another method that Barry and I discovered by accident: I had cut some flowering blossoms to take indoors, and discovered too many ants on them. So I put the stems in a vase of water and left them outside for the night. The next morning, the ants had disappeared. Barry guessed they had gone home, and I agree.

• Squirrels: Put a little cayenne pepper in a paper bag and shake flower bulbs in it before planting. The squirrels have never dug up my bulbs treated this way. Or sprinkle blood meal over your bulb bed when the shoots come up. To keep squirrels off your bird feeder, use an inverted cone guard on the pole. Or look for a bird feeder that has a mechanical trip on the door when something heavy sits on the ledge.

• Rabbits, groundhogs and raccoons: one inch galvanized wire mesh mounted three or four feet high on metal stakes will keep these pests out of the garden. Leave the top 18" of the mesh unattached, so it will be floppy for any animal that tries to climb it. Bury the bottom 12" of wire underground. For a city garden, leave the cat out at night. Although I doubt that our cat Goldwyn could best the giant raccoons in our neighborhood, they seem to have maintained some sort of truce, and the raccoons seldom disturb anything in our yard.

Keep the garden clean.

Sanitation will help control many insects and diseases. For example, picking up all the fallen leaves from black spot infested roses helps prevent the spread of black spot through the soil.

Vary your crops as much as possible. Plant pest and disease-resistant varieties.

Rotate your crops. Grow your tomatoes in a different plot, or plant something else for a couple of years if you have only a small plot. Alternate your root crops such as potatoes or carrots with foliage crops such as lettuce or spinach. Or alternate fruiting crops, such as tomatoes and squashes with leafy crops like swiss chard. That way, pests and diseases which affect any given crop will not have a host plant for the following season.

Do not compost severely diseased plant material unless you can bury it deeply in a cooking compost pile, or avoid using the compost for a couple of years.

• Black spot on roses: A fungus, black spot really loves damp weather. Tomato leaves contain solanine, a chemical that inhibits black spot disease. Save your tomato leaves when you're pruning. Grind up two cups full of them, and mix them into a half gallon of water. Stir in a couple of teaspoons of cornstarch or buttermilk to help it stick. You can keep this mixture in the fridge, and spray it on as necessary.

The most important control on leaf spot diseases is sanitation. Pick up diseased leaves and burn them in your barbecue or put them in your tightly sealed garbage.

These diseased leaves are one of the few things from the garden that I don't compost. Be sure to water roses with a slow soak at ground level. Don't sprinkle them from above, as wet leaves are more susceptible to black spot.

Experiment with companion planting.

Companion planting has proved popular with many organic gardeners. The idea is to plant together species which will benefit each other, and help prevent disease and insect infestation without chemicals. I have had limited success with companion planting, but as Barry says, you don't know how bad the problem would have been if you didn't try to solve it. I've never had the luxury of enough space to plant a "control" crop along with an experimental crop.

My recent experiment of planting tomatoes among the roses to control black spot led to mixed results. The tomato plants were so vigorous from all the compost in the bed that they grew high, wide and handsome, and became totally entangled in the rose thorns. The tomatoes were terrific, but I think that the extra foliage around the roses prevented good air circulation, which may have partially foiled the attempt to control the fungus.

Other well-known companions include garlic, which will repel many insect species. Onions or chives may help control rust fly on carrots. French marigolds are said to be beneficial to tomatoes, potatoes and beans. Mint, thyme or sage may repel cabbage butterflies. Plant tomatoes near asparagus to discourage asparagus beetle. Basil among the tomatoes may deter tomato hornworm. I am somewhat tentative in offering these solutions. They may work in some situations, or only partially work in others. If you're interested, by all means experiment with companion planting. It will certainly do no harm.

Avoid chemical fertilizers.

If you feel you have to fertilize, what should you use? A great top dressing for lawns is that wonderful rich crumbly compost that you're making all year long. Sprinkle it on your lawn. Dig it into flower and vegetable beds. Some fertilizers advertised as environment-friendly and relatively harmless to soil micro-organisms, lakes and rivers have been attacked for being less than perfect. My feeling is that very little is perfect in this world, so if you plan to fertilize, choose as friendly a fertilizer as you can. Check your hardware or garden center for other organic fertilizers.

Wood ashes, rock phosphate or lime, bonemeal, manure and bloodmeal are all natural soil enhancers. Your garden supply center will carry these and other organic additives. Mulches from grass clippings, newspapers, hay, leaves, even brown paper bags, are all beneficial to your garden, and will help retain heat and moisture. An aluminum foil mulch will repel aphids. Rake mulches away from plants that suffer from fungus disease.

> **Make sure your lawn care company has chemical-free maintenance. If they don't offer it, check around with competitive companies.**

Some lawn care companies put little warning signs on the lawns after they spray—a sure sign that poisons have been sprayed. But many lawn and garden care companies are beginning to see the light. Many are now offering service based on organic gardening principles, rather than chemical controls.

Look for domestic varieties and hybrids. Ask for the origin of any purchased plants.

Check to make sure that bulbs are from propagated plants, rather than collected from the wilds of countries where environmental protection may be lax. Don't dig up native wildflowers or buy them at market stands to plant at home. They may not thrive in your garden for more than a few years, which could contribute to the extinction of endangered species. Avoid buying imported species such as rare dwarf aloes and euphorbias from Madagascar. Avoid "hardy" cyclamen bulbs, *Tulipa praecox* (other tulips are okay), giant snowdrops, scilla, and yellow-flowered winter aconites.

Buy organically grown produce.

Support organic growers. If your local supermarket or natural food store offers certified organic produce, try it at least once. If they don't have it, ask for it. Sometimes it will be more expensive than mass produced commercially grown produce. But the more we support organic growing, the more the stores will stock it, and the price will soon be competitive. It will eventually mean less chemical production worldwide and increasingly rich agricultural land, as opposed to increasingly depleted land. You may find that you like it so much you will decide to buy only organic produce. Besides, isn't your health worth a few cents more?

Plant a tree this weekend.

If you have space on your property, at your business, or at your school, plant a tree. Trees are a vital component in the health of our planet. They filter pollution from the air we breathe, contribute oxygen to our planet, secure the soil from erosion, shelter birds, insects and other animals, help to control the greenhouse effect, act as air conditioners, cooling our air in the summer time, and shedding their leaves in the winter so the sun can warm the earth.

Donate a pussy willow to a park, an oak tree to a school, a spruce tree to a hospital or a cherry tree to your city or town in memory of someone. Give a linden as a graduation gift, or a ginkgo tree for a housewarming present, and offer to help plant it. When someone asks you what you want for your birthday, tell them a Scotch pine or an apple tree.

One of the nicest presents I've ever received was a beautiful maple tree from our parents when Barry and I bought our first house. Our neighbor was a retired gardener with a superb garden. He told me it was the first time he'd ever seen a woman plant a tree. (I told him it wouldn't be the last.) Although I loved that house dearly, the thing I most miss is the maple tree.

I often think with gratitude of the person, whoever it was, more than 60 years ago, who planted the huge sugar maple that now towers over the third story of our present house. Our maple tree cools our house so well that I can work in my third floor office without an air conditioner, even in the worst heat waves of summer.

It also provides a resting place for cardinals, jays, robins and crows, a runway and playground for black, brown and gray squirrels, a shady canopy for our front bedrooms, and a quiet, inviting presence on our street.

> *The more we come in contact with the planet Earth—air, soil and water—the more we learn to respect the diversity of nature. "Speak to the earth, and it shall teach thee."—Job XII, 8.*

13

THAT LITTLE GREEN COTTAGE IN THE WOODS

And Other Green Recreation Tips

Humans don't bother to ask when they want to invade any corner of the Earth; we just barge right in wherever we feel like going. This has been called, in the past, adventuring, pioneering, trailblazing, and other hearty sounding recreational names.

I'm all for camping. I love to get out in the wilderness and see the stars blazing brightly like we never see in the city. I love going to sleep with absolutely no human sound in my ears, just the sounds of the night wilderness. I think it's important that everyone experience nature up close as often as possible.

My sister Frances was smart and lucky enough to marry a cottage owner. And my brother-in-law Tiit has been generous in welcoming us all as guests. While there, I've learned many cottage owners and campers are beginning to complain that the lakes are being polluted, the forests are dying, and the neighbors are noisy to boot. The bad news is that in the wilderness it's hard to lay the blame on large corporations. The good news is that since individuals are most of the problem, individuals can make a big difference.

Those who are privileged enough to spend time in the wild have a responsibility to keep it as nearly pristine as possible. By pristine, I mean a state of wilderness without human interference.

If we invade the home of moose, beaver and fish, what will they think of the condition in which we leave it? Will they think we're not fit house guests, and hope we never return? The wilderness is not just home to these few identifiable animals, but to thousands of species, some microscopic, some we may never know, which all interact with each other to ensure a rich and teeming ecology. We're privileged guests in their home. Let's not abuse our privileges.

What To Do At The Cottage Or Cabin

> **Take your basin of water 50 yards away from the lake or river for shampooing and for all types of cleaning.**

We've been washing our hair in the lake for years, but this is the year I quit. Even though we use biodegradable shampoo, I can no longer justify putting the shampoo in the water. Our kids swim in the lake, people fish in the lake, birds nest on the shoreline, and plants grow along the beach. Shampoo is not a natural part of the marine ecology. I know I'll probably regret making this commitment in print. It's very convenient to wash hair in the lake. We swim every day, and the shampoo sits on the dock. And I'll probably alienate everyone in the family because they'll think I'm pointing the finger at them and being self-righteous. But if I'm asking you to do this for the environment, then I can't go ahead and do the opposite myself.

If you're already in your bathing suit, two minutes should be enough time to wet your hair in the lake, grab a pail full of water, and stroll into the woods to shampoo and rinse.

Throw out your dishwater away from the lake.

If you wash your dishes in a basin, be sure to buy environment-friendly dish soap. If you can't find any labeled as such, use laundry soap flakes. The flakes might seem difficult to dissolve, especially if you have hard water, but will clean your dishes just as well. Detergents often contain phosphates, enzymes, perfumes and other ingredients, which are harmful to our lakes. Soap flakes will break down very quickly. If you don't have a septic system, give the Earth a chance to break down some of those ingredients before they leach through to the lake water.

Eat an orange to avoid using pesticides and insecticides.

Insects don't like the smell our skins give off when we eat oranges and other citrus fruits. (But avoid bananas— that's one of their favorite flavors.) For every fly or mosquito we kill with insecticide, hundreds more will come to take its place. But every time we use a poison, we risk killing off our natural insect predators, the birds. We also kill all the beneficial insects at the same time as the pesky ones. Then we're left being dependent on more and more pesticides.

Avoid perfume, and scented shampoo, deodorant or creams. All these will attract insects too.

Insects are attracted to darker-colored clothing, so wear light colors for your tramp through the woods.

Inside the cottage, keep all food in sealed glass or plastic containers. Then if you happen to see the odd ant or fly, you won't worry that your food is being contaminated. It also means you can leave your nonperishables in the cupboards while you're away, without fear of either insect contamination or insecticide contamination. Keeps out the mice, too.

And the purple martins won't mind if you put up a home for them here, too.

Do not use anti-fouling paints on boats, docks or fish nets.

Tin-based paints, which are used to preserve boats and docks, usually contain tributyltin, or TBT, a highly toxic chemical. These anti-fouling paints prevent algae and shellfish, such as barnacles and mussels, from clinging to boat hulls and are also used on salmon farm nets, and as a preservative on docks and other marine structures.

Research has shown conclusively that TBT is extremely damaging to marine life, including fish, lobsters, snails, clams, mussels. It's been found in salmon and oysters all over the world. Cooking does not destroy the toxin, so

2 Minutes a Day for a Greener Planet • 205

if we eat contaminated seafood, the poison starts to collect in our bodies too. Little is known about the long-term effects of eating seafood contaminated with TBT, which, according to the National Water Research Institute, "may be the most acutely toxic chemical to aquatic organisms ever deliberately introduced to water."

When Camping

Stick to established campsites.

A campsite creates a disturbed area in the wilderness. Choose an existing site, and do not change its size. If it's not big enough for you, either your group is too big, or you have too much stuff. Groundcover and other vegetation may be trampled and destroyed by careless expansion of campsites.

Use the campsite as it is. Don't try to change it.

Changes to a campsite include most kinds of digging, and any hacking, chopping, piling, sawing, or stacking. Low-impact camping means camping as if we were fugitives, on the run, hoping to leave no trace of where we've been. The object is to make as little impact as possible on the environment, and to try to leave it better than we found it. Every footprint makes an impact on the environment. Try not to widen or lengthen existing trails, nor to shorten them by taking shortcuts.

Limit the number of people for each campsite.

It's better to have small groups rather than large. The more people, the more unwieldy the group, and the more danger of disregarding environmentally sound practice. Six people is about right for any one site. If you happen to be seven friends camping, you'll make less impact with two small campsites than with one large one.

Take nothing live from the wilderness, whether trees, plants, flowers or animals.

Use a self-supporting tent. Don't cut live trees for tent poles. It may look as if the wilderness can stand any amount of hacking away, but if every camper took a few saplings, our wilderness would suffer enormous damage.

Don't strip bark from live trees.

Although it's tempting to strip bark, especially when we see that lovely white birch bark, we can seriously damage the trees. Stripping their bark may or may not kill them, but if we leave them alone, we know for sure that we haven't harmed them. Teach the kids to respect living trees, and to protect them. If anyone wants to write a letter on birch bark, find a piece of it that has already fallen to the ground.

If you must have a fire, use only fallen, dead wood.

A lightweight camp stove is preferable to a fire, but not always practical. Still, some kinds of fires are better for the environment than others. If you're canoeing, gather wood along a mainland shore. In any case, keep your fire small—big enough to cook your meal, but no larger.

Instead of throwing cigarette butts on the ground, put them in one of your empty cans, since you'll be taking the can with you when you leave.

Besides the obvious hazard of a forest fire, the nicotine in cigarette butts is toxic to wildlife that are poisoned by them.

Do not use disposables of any kind.

Use lightweight reusable plastic or metal dishes instead of disposable plates and cutlery. You only need one set per person, and they last for years. Disposable paper or foam plates become garbage after one use. If you're in a controlled campground with garbage cans and dumpsites, the garbage you leave behind becomes the problem of that particular municipality. Better not to create garbage in the first place.

Anything you don't burn, carry out with you.

The best campers leave the wilderness cleaner than they found it. If you spot garbage anywhere along the way, pick it up and take it out with you. It's a good habit to teach our kids, not just for camping, but for everyday life.

If there are no toilets, bury human waste in a shallow pit.

If the campsite has toilet facilities, use them. It's better to use existing outhouses than to leave human waste elsewhere. In the wilderness, dig a small hole, no deeper than six inches, at least 50 yards away from any lakes or rivers. Use single-ply white or unbleached toilet paper, or if you really want to practice low-impact camping, use leaves. (Be sure you know your plant species so you don't choose a poisonous plant.) Cover the waste with the dirt you dug from the hole. The top layer of soil is rich in bacteria and other micro-organisms that will break down the waste quickly. Since human waste is compostable, it will eventually become part of the forest soil.

Any paper-based sanitary products may also be treated this way, although it's a good idea to shred them, if possible. Do not bury plastics, including disposable diapers. Plastic tampon applicators and plastic napkin liners should be carried out with your garbage.

Choose soap over detergent.

You don't need to haul detergents and shampoos into the wilderness. I use a bar of glycerine soap for face, hair, dishes and underwear. Use a basin instead of the lake or river for washing and rinsing, and dump your waste water at least 50 yards away from the open water.

When Hunting Or Fishing

Avoid lead sinkers and lead shot.

Waterfowl eat both lead sinkers (which may be lost from fishing lines) and lead shot, and may die an excruciating death by lead poisoning. The lead paralyzes the digestive tract, and the bird dies a long, slow death after 17 to 21 days.

Several U.S. states have already passed laws requiring that steel shot be used for hunting, and Britain has converted to steel sinkers for angling. If you are traveling to a country where steel sinkers are available, pick some up as a great green gift for a fishing friend.

> The poet Christina Rossetti wrote:
> "One day in the country
> Is worth a month in the town."
> Haven't we all felt that way, at least once?

14

GREEN PLANET PETS AND WILDLIFE

The Chain That Binds

What's the difference between us and the rest of the animal world? This age-old question has some interesting new answers. We used to define ourselves by our intelligence, our memory, our foresight, or our religion.

Now we have a new criterion by which to measure ourselves against the rest of the animal world: our effect on the environment. Left to their own devices, animals could not in 10 million years of evolution come up with acid rain, ozone depletion, excessive packaging, the greenhouse effect, disposable diapers, carbon monoxide poisoning, nuclear bombs, toxic dumpsites, DDT, or leaded gasoline. We've done it all in less than a century.

Human begins are wiping out species at the rate of a thousand or more every year.

Yet we love animals. We take them into our homes, feed them, pamper them, treat them as family members. Our pets encompass a wide cross section of species. We adopt pigs, snakes, spiders, fish, horses, frogs and rats, as well as the more common cats, dogs and birds.

We're not reluctant to spend money on our pets. For some pet lovers, nothing is too good for Fido or Fluffy.

The British spend $3.7 billion per year on their pets.

The pet food industry consumes tremendous quantities of proteins, requires land, machinery, energy and money, all of which might be put toward other uses. But none of that convinces Caroline when she's playing with our cat Goldwyn. The bond between humans and animals is one link in a chain that binds all living things. Our domestic animals are a real and tangible connection to nature and the ecology of our planet. If we can love a dog, cat, spider or snake, we're a step closer to grasping the responsibility human beings have for the rest of the living world. Once we grasp that kinship, we may begin to behave in a manner that respects all forms of life on Earth.

What To Do

Look for biodegradable cat litter.

There are several available, at least one of which is made from recycled newspaper. Use baking soda to control odor. Avoid throw-away liners. They're unnecessary. Washing the litter box regularly will keep it fresh.

Make your pet door energy efficient.

Does your pet have its own entry/exit door? That pet door can let a lot of heat out of your house, and waste a great deal of energy. One solution is the double pet door. Put one pet door in the outside door, and another in the inside door. Stagger the position of the doors, so the draft won't blow right through. Don't worry—your pet will learn very quickly how to go through two doors to get in or out.

A seal of weather stripping around the outside edges of the pet door will greatly increase its energy efficiency. Or insulate the pet door by hanging a small decorative quilt or doll blanket, completely covering the outside edges of the opening. Save heat and be more comfortable.

Put a sticker or decal on your windows to help prevent birds from hitting them.

Thousands of birds die every year from hitting windows. No, it's not suicide. They tend to focus on the reflected image in the window, rather than the surface of the window. Check your windows from outside to see if you can see reflections. If you can, birds can. Fasten a decal or cutout silhouette of a bird on the outside of the glass, giving the bird something on which to focus at the surface of the window.

If you do find a bird which is stunned from hitting a window, put it gently in a box, and keep it warm until it has recovered. Do not try to feed it. If the bird has not recovered after a few hours, a veterinarian may be able to help.

Put up a bird house.

Build or buy bird houses, including purple martin houses. Purple martins are almost totally dependent on human beings for their dwelling places now, and are an excellent control of mosquitoes and other small insects. See Chapter 12 for more on purple martins and other bird houses. They're also great for teaching kids bird-watching and can supply entertainment all year.

Try non-toxic flea remedies.

If your pet goes outdoors, it will almost certainly come into contact with fleas. Avoid toxic flea remedies as much as possible. There are powders and sprays and mousses, but all of them have poisonous or toxic ingredients.

The manufacturers admit it. A warning on a flea and tick powder container reads:

> **"Do not reuse empty container. Do not contaminate water by disposal of waste."**

Specialists on hazardous waste suggest that flea powders be fully used and the containers triple rinsed, reusing the rinse water as a flea killer itself. Then the container could be disposed of at a municipal landfill. All of this suggests to me that flea powders are extremely toxic, and best avoided altogether.

Mature fleas feed exclusively on the blood of birds and mammals (that includes us). They're notoriously agile jumpers, and, with their hard bodies, difficult to kill if you can catch them. If you catch one between your fingers, squeeze very hard, preferably with a fingernail—otherwise, they'll just jump away again when you let go.

The best weapon is destruction of the immature stages—the eggs, larvae and pupae. These will be found amongst the pet's bedding or in the areas the pet frequents. Vacuuming will help to control infestation effectively. Vacuum religiously, especially in carpeted and upholstered areas, and the pet's sleeping area. Empty your vacuum into the hot center of your compost, or wrap tightly before discarding. Do it often so the fleas don't just set up housekeeping inside the vacuum. If your cat or dog has a sleeping basket, line the basket with newspaper which can be discarded—or use an old towel, and wash it regularly.

Brewer's yeast, if you can get your pet to eat it, may help control fleas. It's available at all natural food stores. Try ½ tablespoon brewer's yeast in your pet's food every day. Some cats seem to love the taste of it. If your dog has no objection, try garlic powder in its food. Avon produces a bath oil called "Skin So Soft," which may help to repel fleas on both humans and pets. Try it in your dog's rinse water after its bath. Check your natural food stores for organic flea collars—or make your own terry-cloth collar with a Velcro fastener. Soak your homemade collar for ten minutes in boiling water with a handful of rosemary. Let cool and dry, and fasten on your pet. Use the rest of the water, after it has cooled, as an after-bath rinse for your pet. Let your pet dry indoors without toweling.

Another rinse that may be useful, is simply salt water. Be careful of cuts or scratches on the animal's skin, and avoid eyes and genital areas. Consistent brushing of your cat or dog helps too. I brush our long-haired black cat outdoors before letting him into the house.

For a commercial flea product low in toxins, try Safer's Flea Shampoo or Safer's Flea and Tick Spray, available at pet stores, garden centers, or from Echo-Logic—phone (416) 360-8799. Safer's Flea Shampoo contains potassium salts of naturally occurring fatty acids. Their spray contains pyrethrum, an organic insecticide made from chrysanthemums. It has a low toxicity, but is effective on contact with fleas and ticks. The first time Barry bathed Goldwyn with flea shampoo, he howled pitifully (Goldwyn, not Barry). But his coat came out fluffy and shiny, and he soon forgave us.

Poop goes in the toilet, not the trash.

If possible, dispose of animal waste in a toilet, or compost it. That includes cat waste scooped from the litter box. I know it's easier to chuck the bag in the nearest trash can when you're out in the park, but that trash goes out to our landfill sites.

Compost the droppings from your own animal.

Cat and dog feces are compostable, but don't use that particular compost pile for your vegetable gardens, only for flower gardens or lawns. Same goes for cat litter. Guinea pig, hamster, rabbit and bird droppings are all compostable. You may already be buying pig or sheep or cattle manure to fertilize your garden. If you have a large dog, you could make a significant pile of manure from its waste. Cover each layer with a shovel full of dirt to keep it sweet smelling and to break it down quickly. See Chapter 11 for more on composting pet manure.

Compost your pet's hair, fur or feathers.

Does your collie shed, your budgie molt, or your rabbit pull out its fur? Who would have thought all those leavings are compostable—even from your vacuum cleaner? Even thrown up hairballs can go in the compost. Never mind saying, "Ewww, yuck!" You have to clean them up and discard them somewhere, don't you? I've never had the opportunity to compost a discarded snake skin, but there's no reason not to. All material made from living cells is organic, and is a candidate for composting.

Choose domestic pets.

Don't buy exotic pets. Despite attempts at regulation, tropical and foreign species are often imported illegally. These animals, birds, or fish may survive for a few months or years, and then die because of improper diet, care or climate. They may not have the opportunity to reproduce, and may become scarce, endangered, or even extinct. Estimates suggest that for every tropical bird smuggled successfully into the country, 10 may die from mishandling and efforts to conceal them. Because they are not quarantined when they are smuggled into the country, the ones that do survive may spread highly contagious diseases to domestic birds or animals.

If you choose any unusual pet, especially reptiles or birds, make sure that it has been bred in captivity, and not captured from the wild. And please don't give pets as gifts without checking first with the recipient that such a gift would be welcome. A pet is a lifetime commitment to care.

Protect our world's elephant population by boycotting the purchase and sale of ivory.

Every year, we kill from 60,000 to 100,000 elephants in a seemingly insatiable demand for ivory. Don't buy ivory, or any products containing ivory. Although there is a small legal trade in ivory, a great deal of it is illegal, and it's often impossible to tell which is which. Ivory is carved to disguise its origins, and the illegal ivory becomes mixed with the legal. Elephants are being slaughtered to the point of endangerment.

In Africa, 2000 elephants are killed every week. In 1979, there were 1,500,000 elephants in the world. Now there are fewer than half that many. Sadly, very few elephants live to maturity any more. Years ago, tusks weighed 100–150 pounds. Now most tusks weigh less than 10 pounds because poachers kill younger and younger elephants as mature ones become more scarce.

Be especially wary if traveling in the Far East, where ivory abounds. Both Hong Kong and China, two of the world's largest users of ivory, have announced that they will defy an international ban on ivory trade. Hong Kong has an 800-ton stockpile of ivory. Besides jewelry, statues, and ornamental carvings, ivory can be found in billiard balls, piano keys and inlaid wooden articles. Yamaha and Kawai, two major Japanese piano makers have pledged to stop importing ivory for piano keys. Check with your piano dealer and manufacturer before making a purchase. New plastic keys simulate ivory. Despite my preference for natural materials over plastic, this is one time I'd choose the plastic.

Some jewelers have voluntarily withdrawn ivory from their stores recently. Ask yours to do the same. Don't forget about small items like dice, chopsticks, prayer beads, trinkets, toothpicks, souvenirs, or game tiles. If consumers don't buy ivory, the poaching will stop.

Boycott the purchase and sale of products made from wildlife.

Avoid buying tortoiseshell, wildlife skins, furs, coral, and shells. Watch out for carvings and jewelry. Even sea rocks that seem safe can be a problem. Coral reefs are hacked away for souvenirs, even though they are a refuge for many sea species.

Choose regular balloons for your next party, not helium balloons.

We've all wondered where helium balloons go when we release them into the sky. How high? How far? Unfortunately, many are carried out into the ocean, or to seashores. Every year thousands of animals, birds and fish die from ingesting them. Whales, dolphins, turtles and seabirds eat the balloons, or pieces of them, mistaking them for food. Sea turtles mistake balloons for jellyfish. The latex blocks the animal's digestive system, preventing any further food from being absorbed. The birds or animals usually die a lingering death.

Buy singles or boxes of canned drinks.

Don't buy canned drinks connected with plastic six-pack rings. Plastic six-pack rings have been directly responsible for the deaths of numerous birds and animals. Birds, in particular, become entangled in the rings, and are strangled to death. Other birds and animals may starve to death when the rings become stuck on their beaks or snouts.

More than 100 million pounds of plastic are dumped into our oceans every year, according to the World Society for the Protection of Animals.

Even though we ourselves may be careful with our plastic garbage, we have little control over what happens to it when it leaves our curbside, or when we dump it in a public litter barrel. If we continue to buy plastic six-pack rings, manufacturers will continue to make them, and others who are not so careful as we are will buy them. If you are at a gathering where plastic six-pack rings are in evidence, cut them apart with scissors before throwing them away.

Demand "dolphin safe" tuna.

Be careful when buying canned tuna. You could be endangering dolphins. According to Greenpeace, yellowfin tuna associate with dolphins in the southeast Pacific. Tuna fleets set mile-long drift nets near dolphins in order to capture the tuna. As many as 100,000 dolphins per year may die from being caught in tuna nets. Previously, we have been urged to buy only canned tuna marked as being albacore or tongol, since these varieties are less likely to be fished from dolphin habitats. But now even those varieties are suspect. The U.S. federal government has introduced a bill which would require tuna products to be labeled "Dolphin Safe." Support the bill by writing to Rep. Barbara Boxer, U.S. House of Representatives, Washington, DC, 20515.

One day a month, go without meat.

Many North Americans are eating less meat than we used to, partly for our health, and partly out of the knowledge that meat consumption wastes our planet's resources. Frances Moore Lappé, in her wonderful book, *Diet For a Small Planet* (Ballantine Books), documents the hideous waste of protein fed to livestock compared to the minuscule amount of protein we receive from livestock in return. We could easily supply the human population of the Earth with enough protein if we stopped feeding it to our livestock.

Cattle consume more than 15 pounds of grains for every pound of beef they give us in return.

The demand for beef also means the clearing of vast tracts of tropical rainforest for cattle grazing. The land rapidly deteriorates, the soil erodes and becomes infertile. Then more acreage must be cleared for cattle grazing. Land which once supported farmers in tropical countries now grows soya—not to feed the people, but for export as livestock feed. Some fast food chains get their beef from tropical lands such as Costa Rica. Ask your fast food outlet where their beef comes from.

As cattle digest, they give off methane, a greenhouse gas. The world's cattle population, along with large areas of rice paddies, account for nearly half the global release of methane. More beef means more global warming.

Finally, livestock grazing requires tremendous amounts of water to irrigate pasture land. California alone uses enough water to meet the domestic needs of 22 million people, just to turn the desert into grassland for cattle and sheep grazing.

If we all reduced our meat consumption, we'd make a significant impact on the protein available for the rest of the world, save water for more reasonable uses and help preserve our tropical rainforests, which we desperately need for the health and survival of the planet.

> *Kinship with animals marks the start of our kinship with the rest of the living world. We depend upon the diversity of all forms of life for the survival of our planet.*

15

FESTIVITIES AND GIFTS

The Gift Of The Green

For many people, good intentions fall to pieces in the face of tradition. "That's the way we've always done it," seems a reasonable excuse for continuing to do things the same old way, regardless of whether there may be better, more environmentally friendly ways of doing things. Some people of good will and good intentions simply cannot bring themselves to give up things like wrapping paper, or individually wrapped and overpackaged chocolates, because they can't imagine how else to celebrate.

Others might welcome the change but feel reluctant to impose the new ideas on family members, and so go along with family customs, even though they'd like to try something new. I know we're victims of the syndrome in our family. Even though most of us aren't very fond of Brussels sprouts, somehow they've become part of our family Christmas dinner, and we always have them. No one seems willing to throw caution to the wind and say, what the heck, let's have peas this year instead. Admittedly, it's a small example, but a symbol of what we're up against in trying to change traditions.

I think a good way to tackle the situation is to go for one problem at a time, and get the support of family and friends. Soon, they'll be leaping in with suggestions of their own on greening up the festivities.

What To Do?

Rent dishes, napkins, cups and saucers, tablecloths and glasses instead of using disposables.

Having a big office or house party? Most of us don't have on hand 50 or 100 plates, wine glasses, napkins, coffee cups and saucers. So some people resort to disposable everything—paper or foam plates, plastic glasses, paper napkins. Nobody really likes eating and drinking from these disposables, but they're certainly handy. There are whole industries that do nothing but rent party equipment. Besides, who wants to drink champagne from a plastic glass? The great thing is that you don't even have to wash the dishes or glasses. Just send them back dirty. Wine glasses and plates rent for 25¢ to 35¢ each, and napkins from 35¢ to $1.20 each.

Another way to do it is to have one or two friends bring all the plates or glasses they can spare. You know, often the best fun at the party happens in the kitchen when cleaning up.

Use your imagination for gift wrap. Create colorful wrappings from leftovers.

You don't need to buy gift wrap. At those prices? And watch the kids take one look and tear it to shreds? You have all you need, right in your own house.

Famous gift wraps I have known: the colored comics section of the newspaper. This is the best one, and the kids love it. In our house, we save the weekend comics at certain times of the year for all those birthdays that seem to fall in the same month. Old scraps of wallpaper—very sturdy—will last through many wrappings. Egg cartons tied with a bow: I once gave my sister Frances a half dozen pairs of summer socks in an egg carton—one sock for each egg pocket. Reusable home-sewn cloth bags. Old colorful posters. Fabric scraps. Or the famous present within a present: a hat wrapped in a matching scarf, a piece of jewelry in a decorative wooden box, a baby's sleeper wrapped in a receiving blanket, homemade cookies in a reusable tin or cookie jar. All that's required is a bow to tie them up—no paper necessary.

Don't forget to reuse any gift wrap you receive. If you take just a few seconds longer unwrapping gifts, you can do it without destroying the paper. Simply cut off the torn edges and re-fold the paper for the next use. There's a standing joke in our family: "Haven't I seen that paper before?"

Same thing for those decorative ribbons and bows. They'll last through dozens of gifts. We put them in a bag when we unwrap gifts, and then we always have a selection from which to choose.

Choose a cloth or knitted bottle bag.

Taking a bottle of cheer with you? You don't need a fancy paper gift bag—a simple bow will do. But if you insist on having it wrapped, choose a bag that will be reused many times over. You may even be lucky enough to get the same bag back again yourself one day. Knitted bags are terrific because they stretch to fit so many odd sizes of gifts besides bottles. There's a cloth bottle bag that has been making the holiday rounds of my family and friends for a few years. It has a beautiful cloth ribbon stitched right into the seam, so the bow is part of the package.

If you already have an artificial Christmas tree, treat it with care.

The Christmas tree dilemma. As an environmentalist, the choices are daunting. Some suggest using an artificial tree. Many apartment buildings forbid real trees on the grounds that they may be a fire hazard. Artificial trees are made from plastic and metal, and, like all other manufactured goods, will eventually end up in our waste stream. They cannot be recycled, so the idea is to use them as long as possible. When you're done with it, pass it on to the kids starting out in their first apartment, donate it to an institution, or use the branches to make wreathes.

I have no quarrel with the real Christmas tree, although there are many who abhor that choice. Christmas trees are grown as a crop, on land that would otherwise not support agriculture. Every tree that grows on earth contributes to our oxygen supply and absorbs carbon dioxide for as long as it lives. A Christmas tree grower told me that for every tree cut down, 10 are grown. To me, this makes positive ecological sense. After Christmas, trees should be chipped and/or composted. Do not chop down trees in the wilderness. If you must have a cut tree, get it from a tree farm.

Choose a Christmas tree growing in a pot.

An even better solution to the Christmas tree dilemma is the live potted tree. Many nurseries carry evergreens especially for this purpose, although some will not guarantee them. Prices range from $30 to $100 and up, depending on size and variety. Keep them indoors for no more than two weeks to ensure that dormancy is not broken. Take the tree outdoors again after Christmas, and keep it in a sheltered but sunny area until it can be planted in the spring. Protect the base with extra insulation, and plant it as soon as the ground thaws. Check with your nursery for specific instructions on the care of your live tree. If you haven't the room on your property to plant an evergreen each year, consider donating your potted tree to a hospital, a school, a park, or an industrial site that could use some greenery.

An enterprising nurseryman, David Smith, in Cornerbrook, Newfoundland, has come up with a terrific solution to the problem: Rent-A-Pine. He rents out live potted Christmas trees for customers to take home and decorate. They return the tree, the same as any rented item, and he cares for it through the year until he rents it out again the following year. If no one near you is doing the same thing, suggest the idea to your neighborhood nursery.

You could also consider buying a potted Norfolk pine,

which is meant to stay indoors as a year-round plant, and decorate it for the Christmas season.

Give a green gift.

Do you have a wish list? Do you buy gifts for other people? Consider an environment-friendly gift—one that will make a difference, not just in the recipient's life, but in the life of our whole planet. What better way to say "I love you," "I hope you get better soon," "Happy Birthday," "Good luck," "Congratulations," "Bon Voyage," or "Happy Housewarming"?

Instead of the latest plastic gimmick, consider a gift that will benefit the planet, make the recipient think green, and still be useful and fun.

• For someone who sleeps alone: a hot water bottle or some warm pajamas—keep that thermostat turned down—save energy. Hot water bottles are under $15 at the drug store.

• For the handyperson, homeowner: an automatic thermostat control. Then they'll never have to worry about turning down the heat at night to save energy—it will happen automatically every night. About $50 at any hardware. Easy to install.

• For the athlete: a water-saving showerhead. These showerheads save 50% to 75% of the water, but still give a great shower. They'll also cut down on your water heating bills, which is the biggest energy expense next to heating your home or apartment. Any klutz can install one. They just screw on. About $20 to $70.

• For your boss: membership in an environmental group—try Pollution Probe or Greenpeace, or Friends of the Earth. See Chapter 16 for more ideas. $20 to $30 range—less for students. If you want to be less dramatic, how about a solar powered calculator? About $15 and up. Never buy batteries again.

• For the newlyweds: reusable food storage containers—they could last longer than the marriage. Never buy plastic wrap again.

• For anyone who shops: a cloth or net shopping bag. We're swamped under tons of plastic grocery bags on this planet. Many places, including some supermarkets, are starting to sell cloth bags. $4 to $16. Or make your gift shopping bag from sturdy fabric, and make an extra one for yourself while you're at it.

• For the lucky one who washes the family car: a nozzle end shutoff valve for the hose. They'll save water while they apply the soap and elbow grease.

• For the woodworker: a pattern for a purple martin house. Available free from the Federation of Ontario Naturalists, 355 Lesmill Road, Don Mills, Ontario, M3B 2W8; phone (416) 444-8419. Send a self-addressed, stamped envelope. Or send away ahead of time, and make one yourself as a gift. Great for cottage owners. Purple martins consume thousands of mosquitoes. Never spray pesticides again.

• For Uncle Elmo and Aunt Tillie: a stuffed cloth snake—see if they can figure out what it's for (but don't keep them guessing too long). The long tube of sand-filled or fabric-filled cotton will keep the heat in and the draft out along the crack under the door. Save energy and be more comfortable. Available at craft shows and gift shops for $8 to $12. Also easy to make.

• For your co-worker: a coffee cup just for the office.

Banish foam cups forever. Even though foam cups may be made from so-called environment-friendly plastic foam, that is without CFCs, they're still plastic junk that will be in our landfill sites forever. And they're totally unnecessary.

• For the new graduate: shares in an environment-friendly investment fund. These ethical funds have high standards for investing, and do not put money into companies which pollute or contribute to the destruction of the planet. Check with your stock broker or mutual fund dealer. Some Canadian ethical funds include Crown Commitment Fund, Environmental Investment All-Canadian Fund, Environmental Investment International Fund, Ethical Growth Fund, and Investors Summa Fund. The United States offers several funds, including Dreyfus Third Century Fund, New Alternatives Fund, and Pax World Fund.

• For new parents: enrollment in a diaper service. Diaper service is wonderful. It comes right to the door, takes away dirty diapers, and delivers clean ones. Since an average baby may use up to 10,000 diapers by the time she or he is toilet trained, this will be one gift that will go on giving for a long time. The cost runs around $14 to $20 per week. A diaper pail may or may not be included in the cost. Give as many weeks as you like to get them started. They'll like it so well they'll continue on their own.

• For your dad, who already has everything: buy an acre of rainforest for $25, and save it forever from destruction. Phone the World Wildlife Fund at (416) 923-8173. Use Visa, Mastercard, check or money order. You will receive a decal, an information package, and a blank certificate which you can fill in, certifying the recipient as a Guardian of the Amazon.

The $25 is the approximate cost of protecting one acre of rainforest. Funds go to educate locals on sustainable uses of the forest, such as rubber tapping, harvesting fruits and nuts, and documenting medicinal uses. Nature reserves are fenced and wardens paid to patrol sensitive areas. We cannot survive as a planet without our rainforests. It's a gift that will be remembered long after the neckties are forgotten.

• For the mechanic: re-refined motor oil. Look for the Ecologo symbol of three doves intertwined on the oil package.

• For the young person in a first apartment: "green" cleaning materials, including laundry soap flakes, borax, washing soap, baking soda and vinegar, along with an offer to teach her or him how to use them.

• For the gardener: a subscription to an organic gardening magazine.

• For the cook: a pressure cooker. Save time, save energy, save vitamins.

• To take on a social visit: Brazil nuts or cashews. Widely available around Christmas time, they are products of the tropical rainforest that require a living forest. The harvest of tropical nuts provides a sustainable means of living for locals, which will ensure continued protection of the forest.

• For book lovers: any of a whole stack of wonderful books about the Earth or the environment.

Gifts On The Environmental Blacklist

• Exotic pets: these include many types of birds and reptiles. Many are smuggled in illegally from the Amazon and other tropical countries. Many are rare or endangered. About 10 die for every one that survives smuggling. Those that survive can transmit contagious diseases to domestic animals. Also avoid giving a domestic pet unless you're positive it's wanted and welcomed.

• Ivory of any kind: jewelry, pianos with ivory keys, sculptures or carvings. We've lost half of the world's elephants in the last 10 years to illegal poaching, simply for the ivory trade. It's not worth it.

• Tropical hardwoods: little boxes, bowls, or carvings made from teak, mahogany, satinwood, rosewood or liana. Save our tropical rainforests. Many of the tropical hardwoods grow singly amidst other varieties of tree, which are hacked down, or bulldozed aside to get at the desired wood.

• Electrical gadgets: does the world really need another electric pencil sharpener or can opener? Can we live without lint shavers? If we're managing just fine with the manual tools we have, let's not make ourselves even more dependent on electricity.

• Plastic gadgets: plastic is made from a non-renewable resource, petroleum, which could serve our world much better as a source of energy rather than a source of useless trinkets whose novelty lasts a few minutes at best.

• Disposables of any kind: throwaway cameras are a

foolish waste of energy and materials. Beware toys or gadgets requiring batteries—always buy rechargeable batteries, and if necessary, the recharger to go with them.

> *It's a privilege and joy to celebrate happy occasions with loved ones. I like to think of these celebrations as a continuum of family ties throughout the centuries. Let's ensure that our children and grandchildren and their grandchildren will have a green planet on which to celebrate.*

16

PUT YOUR MONEY WHERE YOUR MOUTH IS

A Little Green Goes A Long Way

Fifteen years ago there were only a handful of environmental groups. Now there are so many that there are some whose sole purpose is to keep track of the others.

I think we owe a big debt of thanks to those early pioneers. Without them the environmental movement would be years behind today. We've come a long way, but we've got some hard work ahead to set the planet straight. And as necessary as individual action is, we still need the power of groups. Government and big business pay attention when a group says, "We've got 50,000 or 100,000 or more supporters who want to see some action."

Most environmental groups are founded and managed by people of good will. Often their salaries are low, their offices modest, and their overheads kept reasonable with the enthusiastic contributions of numerous volunteers. I've been inside many of these offices, and witnessed the dedicated staffers and volunteers answering phones, writing letters, and handling requests. Frills are few, and luxuries non-existent.

But they are not all the same, and not every group is suitable for everyone. Some are more flamboyant, more political, more relentless. Others take a more conservative approach. Most offer some tangible benefit such as a magazine or newsletter to keep members informed. Some offer tax receipts. All are worthy of consideration.

International Groups

Friends of the Earth

530-7th St. S.E.
Washington, D.C. 20009
(202) 543-4312

701-251 Laurier Street West
Ottawa, Ont. K1P 5J6
(613) 230-3352

This hard-working group campaigns internationally with 36 sister organizations worldwide. Concentrating on toxics in the environment, ozone-layer protection, energy conservation, pesticides and public information, they provide well-documented answers to the "What can I do?" questions from the public. Membership fees include a quarterly newsletter.

Greenpeace

1432 U St. N.W.
Washington, D.C. 20009
(202) 462-1177

1726 Commercial Drive
Vancouver, B.C. V5N 4A3
(604)253-7701

6th floor, 185 Spandina
Avenue
Toronto, Ont. M5T 2C5
(416) 345-8408

Greenpeace campaigns for international legislation in 18
countries to regulate and protect the environment. Radical activists on the political left, they're seen by some as
saviors of our planet. They've stopped nuclear testing off
Amchitka, and stopped dumping of radioactive wastes
worldwide. Willing to engage in civil disobedience, they
maintain a high profile in the media. They support bar.s
on commercial whaling. Membership fees include *Greenpeace Magazine* every two months.

Global Greenhouse Network
1130-17th St. N.W.
Washington, D.C. 20036
(202) 466-2823

This network of public interest organizations and legis-
lators represents 35 countries. Alarmed by the trend toward global warming, they hope to generate public
awareness and initiate action on the greenhouse effect.
They monitor global climate changes and disseminate
information.

Nature Conservancy
400-1815 Lynn Street
Arlington, Virginia 22209
(703) 841-8737

The Nature Conservancy works with private land owners to preserve lands all over the world for nature sanctuaries. A nonprofit organization of conservationists, they have special interest in wetlands, forests prairies. They also work to preserve tropical rainforests.

Rainforest Action Network
Ste. A, 301 Broadway
San Francisco, California 94133
(415) 398-4404

The goal of Rainforest Action Network is to save tropical rainforests from destruction for the betterment of the whole planet. They work with other international groups to stage boycotts and public protests, but also organize scientific meetings for public and media education.

Sierra Club

730 Polk Street
San Francisco, CA
94109
(415) 776-2211

2316 Queen Street East
Toronto, Ont.
M4E 1G8
(416) 698-8446

314-620 View Street
Victoria, B.C.
V8W 1J6
(604) 386-5255

The Sierra Club lobbies for preservation of wilderness areas wildlife protection, parks planning, clean air and clean water. They focus on forestry issues, the Great Lakes, Arctic issues, toxic wastes and educational programs. They train members in environmental campaigning. Although sometimes seen as radical, they describe themselves as middle-of-the-road politically. Both local and international excursions are available to members. Membership fees include *Sierra* magazine eight times a year, regular newsletters, discounts on books, posters, calendars and outdoor products, and the opportunity to join outings.

United Nations Environmental Program
North American Office Room DC2-803
New York, New York 10017
(212) 963-8093

The United Nations established this program in 1972 to develop and monitor environmental programs world wide, with particular interest in third world countries. They publish information brochures on a wide variety of topics including global warming, desertification of lands, water distribution and the ecology of the oceans. One of their goals is to provide adequate clean water for all citizens of the world.

World Watch Institute
1776 Massachusetts Avenue N.W.
Washington, D.C. 20036
(202) 452-1999

This is a nonprofit think tank that studies and reports on global environmental issues, along with the social and economic implications of environmental activity. Their annual *State of the World* report is read worldwide and carries enormous influence. Both their regular magazine and special reports are widely respected. A seemingly conservative organization with the ear of influential leaders, they often propose surprisingly radical activities.

National Groups: U.S.A. and Canada

Canadian Environmental Network
P.O. Box 1289, Station B
Ottawa, Ont., K1P 5R3
(613) 563-2078

An information-sharing network of more than 1200 environmental groups across Canada, they provide a national link between eight regional environmental networks. They are working on a catalogue of environmental groups and will send a list of key Canadian groups, with addresses, for $6.

Canadian Environmental Law Association
Attn: Kathy Cooper
401-517 College Street
Toronto, Ont., M6G 4A2
(416) 960-2284

C.E.L.A. uses existing laws to enforce the protection of the environment and to advocate environmental law. A free legal advisory clinic for citizens and groups fighting legal environmental causes, they provide a public resource center. Speakers are available. The annual membership fee of $18 includes a monthly newsletter, *The Intervenor*.

Canadian Coalition on Acid Rain
401-112 St.Clair Avenue West
Toronto, Ont., M4V 2Y3
(416) 968-2135

This high-profile group works with business, conservation and recreation groups to reduce sulphur and nitrogen oxide emissions into the atmosphere, in both Canada and the United States. They have been active in pressuring both government and industry to address the problem. The source of much public information, they are an effective voice for environmental action.

Canadian Nature Federation
453 Sussex Drive
Ottawa, Ont., K1N 6Z4
(613) 238-6154

Originally the Canadian Audubon Society, the C.N.F. promotes conservation of environment and natural ecosystems. Programs include environmental education, parks and protected areas, protection of endangered species, sustainable development and interests in forestry, wildlife and fisheries management. Representing 135 nature groups across Canada, they call themselves "slightly left of center" politically. The $25 yearly membership includes *Nature Canada* magazine four times a year. Call toll-free 1-(800) 267-4088 for their free Nature Canada Bookshop catalogue, which includes outdoor supplies.

Canadian Organic Growers

Membership:	Information: Mary Perlmutter
Box 6408, Station J	348 Briar Hill Avenue
Ottawa, Ont.	Toronto, Ont.
K2A 3Y6	M4R 1J2
	(416) 485-3534

Heritage Seed Program
Heather Apple
R.R. 3 Uxbridge, Ont.
L0C 1K0

The C.O.G. provides information on organic techniques, promotes long term soil fertility, assists in marketing and distribution of organic food and compiles regional directories of growers. Write for the list of books available in their lending library. *A Directory of Organic Agriculture in Canada* is available for $10. The annual fee of $16 ($12 for seniors and students) includes the newsletter, *COGnition*. The Heritage Seed program aims to preserve the genetic diversity of foods (send stamped self-addressed business-size envelope for information).

Clean Water Action Project

317 Pennsylvania Avenue S.E.
Washington, D.C. 20003
(202) 547-1196

They organize both rural and urban Americans to exert public pressure for clean water. They lobby on a wide variety of water-related concerns, including toxics and pollutants, as well as water distribution. They aren't afraid to act politically, and their electoral operations are an effective influence with state politicians.

Defenders of Wildlife
1244-19th St. N.W.
Washington, D.C. 20036
(202) 659-9510

A large national organization (with 80,000 members), Defenders of Wildlife dedicates their energy to preserving native animal and plant species, and restoring and conserving threatened wildlife populations and their habitat. They promote public education and wildlife appreciation. Membership fees include the bi-monthly magazine, *Defenders*.

Earth First!
Box 5871
Tucson, Arizona 85703

These are the hardline activists who first drove spikes into trees they wanted to save, hoping that the risk of damage to sawmill equipment and fear of personal injury would ward off loggers. Not above flamboyant silliness, the Earth Firsters have also been known to dress up in bear suits to protest destruction of forest and wilderness areas. At the extreme left of environmental groups, they have been watched by the F.B.I. Some members have been arrested for illegal activities. This is a group for those who want even more attention than Greenpeace provides. Membership is $20 per year and includes eight issues of *Earth First!* magazine.

Energy Probe
225 Brunswick Avenue
Toronto, Ont., M5S 2M6
(416) 978-7014

A thoroughly reliable source of information on energy related issues, conservation and renewable resources, Energy Probe commands respect in environmental and government circles. Meticulous research has assured them of a loyal following. Always available to the media for comment, they also provide speakers to promote benefits of conservation.

Federation of Ontario Naturalists
355 Lesmill Road
Don Mills, Ont. M3B 2W8
(416) 444-8419

One of the oldest environmental groups (since 1931) and one of the most effective in lobbying governments, the F.O.N. created the Coalition on Niagara, Conservation Authorities, and dozens of new parks. They are active in protecting and increasing awareness of wildlife, and offer excellent education programs for children and seniors. Their annual fee of $28 includes the quarterly magazine *Seasons*, the opportunity to join nature trips and a catalogue of nature-related items.

Great Lakes United
Attn: David Miller
1300 Elmwood Avenue, Cassety Hall
Buffalo, New York 14222

A joint U.S.–Canada organization, Great Lakes United is aimed at the protection and preservation of the Great Lakes and St. Lawrence River. This coalition of over 200 member groups includes environmental, business and government interests.

National Audubon Society
950 Third Avenue
New York, New York 10022
(212) 832-3200

Widely known and respected for their glossy publications, *American Birds and Audubon,* the National Audubon Society boasts a membership of more than a half million. A conservative approach has kept them in business since 1905. Although named for the famous painter of birds, James Audubon, the Society's interests encompass other fields of conservation. They operate 250,000 acres of wildlife sanctuaries.

National Wildlife Federation
1412-16th St. N.W.
Washington, D.C. 20036
(202) 637-3700

Larger than most of the other environmental groups put together, the National Wildlife Federation claims 4.8 million supporters. They carry a great deal of weight with Washington politicians, who trust them to deliver their message without heavy-handed tactics or excessive finger-pointing. Their aim is to preserve wildlife and promote conservation. If you like to be part of a large crowd, this is the environmental group for you.

Natural Resources Defense Council
122 East 42nd St., 45th floor
New York, New York 10168
(212) 949-0049

Famous for their association with actress Meryl Streep,
the National Resources Defense Council struck a public
nerve when they publicized the widespread use of the
chemical Alar in apples. Much of the public information
on pesticides comes from the N.R.D.C. Their mandate is
to protect the environment through legal action, scientific
research and public education.

Ontario Environmental Network
2nd Floor, 456 Spadina Avenue
Toronto, Ont., M5T 2G8
(416) 925-1322

A nonprofit network for environmental groups in Ontario,
the Ontario Environmental Network includes most of the
national groups. It publishes a listing of more than 300
groups, which you can receive for $6 plus $1 postage
and handing. The book also contains lists of printed and
audio-visual resources.

Planet Drum Foundation
Box 31251
San Francisco, California 94131
(415) 285-6556

Planet Drum Foundation is a nonprofit ecological education group. Founder Peter Berg was the first to articulate
the philosophy of bioregionalism, a direct approach to

saving the planet as a whole by restoring and preserving one's own region. The movement is made up of 60 to 70 local groups across North America, and publishes a newspaper called *Raise the Stakes*.

Pollution Probe
12 Madison Avenue
Toronto, Ont., M5R 2S1
(416) 926-1907

Pollution Probe is active in the areas of toxics, water quality, industrial and household waste management, and environmental foreign policy. Politically to the "center," they make information available to the public, business and government. The membership fee of $30 a year, of which $15 is tax creditable, gets you a subscription to *Probe Post* magazine, a 10% discount on Ecology Store items and a 10% discount on their own publications (Mastercard and Visa accepted). About 85% of the funds go directly into projects.

Rocky Mountain Institute
1739 Snowmass Creek Rd.
Old Snowmass, Colorado, 81654
(303) 927-3128

A nonprofit resource policy center, the Rocky Mountain Institute is a think tank concerned with more efficient use of energy, as well as harnessing cost-effective renewable resources such as the sun, wind, and flowing water. The founders, Hunter and Amory Lovins, are responsible for ground-breaking work on the boomerang effect of economic development plans in small towns. They endorse the concept that energy conservation is more vital than military spending to national security.

WEB
2nd Floor, 456 Spadina Ave
Toronto, Ont., M5T 2G8.
(416) 929-0634

WEB is an electronic network of communication services and information retrieval specifically for nonprofit organizations operating in areas of ecological interest. WEB links personal computers through telephone lines with the WEB host computer. People can communicate individually or as group members. Hundreds of organizations are on-line, including most of the big environmental organizations, as well as many local and international interest groups.

Services include electronic mail, conferences and *WebNews*, an alternative to commercial media stories. WEB is also linked to similar networks all over the world.

Membership costs $25, and subscriptions are $180 per year for a single-user ID account. Monthly charges apply for system usage, telecommunications and data storage. Reduced rates are available for low-income organizations and individuals.

World Wildlife Fund Canada
Suite 201, 60 St. Clair Avenue East
Toronto, Ont., M4T 1N5
(416) 923-8173

The goal of this organization is to establish a network of protected areas, parks, ecological reserves and wilderness areas in Canada. Fields of interest include wild animals, plant and habitat conservation, endangered species work and toxicology research.

Their program for protecting acres of tropical rainforest is called "Guardian of the Amazon." A donation of $25 helps preserve about one acre of Amazon rainforest, educate the locals on sustainable use of the forest, aid the Kayapo Indians in demarcating their land and provide fences and wardens to patrol the nature reserves. The donor receives a certificate as a "Guardian of the Amazon," an information kit and a decal. You can use Visa, Mastercard, check or money order.

There are no membership fees and all donors receive the quarterly newsletter, *Working for Wildlife*. All contributions go directly to conservation projects. Administration costs are paid from earned income, while only 8% of their budget goes to administration and fundraising. Two-thirds of their work is done in Canada.

For a more comprehensive listing of specific and localized groups, the *Directory of National Environmental Organizations* is available for $35 from U.S. Environmental Directories, Box 65156, St. Paul, MN, 44165.

> *Make an investment in the environment. It will cost you little in time or money, but the rewards will be great. You'll gain a more positive outlook on our future, and help to create lasting benefits for the planet.*

17

GREEN LETTERS

What I Learned From Frances

My sister Frances, who works for the government, impresses me with her conscientious answers to letters from the public. Although some bureaucrats may send out standard replies, Frances staunchly maintains that letters to politicians are not ignored. She says *every letter counts*. A letter from a student concerned about the lack of environmental policy in the universities prompted her department to begin to formulate one.

We all have different priorities and different ways of expressing ourselves about things we consider important. If you're interested in writing to people in positions of power to try to influence environmental decisions, here are a few suggested letters.

What To Do

Write to a manufacturer.

Do you have a concern about a manufactured product or process? Look for the manufacturer's address on their product, or ask the store owner who stocks the product for the address. Here's a short, two-minute postcard example you might write:

Dear Madam or Sir:
I am writing to express my concern about your packaging. I have bought your product for years, but I notice that recently you have started wrapping the individual parts in plastic. This extra wrapping is not only inconvenient for me, it is also an environmental pollutant. We have enough plastic in the world already and do not need this excess. Plastic is not biodegradable. It will simply fill up our landfill sites and stay there forever. Furthermore, it's a blatant misuse of our non-renewable petroleum resources.

I enclose the excess wrapping for you to deal with. I regret that I will not be buying your product again until I learn that you have stopped using the extra, unwanted plastic wrapping. I await your reply.

Write to your local politician.

If your concerns are local, start with your local politicians. They have the control over your dump sites and incinerators, your water quality, your air quality. City, town and municipal governments are closest to the concerns of individuals and very sensitive to their electorate. What should you say? Simply express your concern over the environmental impact of whatever problem you wish to address.

Dear Madam or Sir:

I am writing to express my concern over our city's policy of allowing recyclable materials to go to our landfill site. During the past few years municipalities everywhere have started recycling newspapers, cans, bottles, corrugated cardboard and plastics. Meanwhile, our community continues to let those reusable natural resources pile up in the city dump and further destroy our beautiful countryside.

I urge you to commence action to get a recycling program started in our community. I await your reply.

Write to your representative in the federal government.

Want to address national or international concerns? Here's a letter you can write about a national problem:

Dear Madam or Sir:

I am writing to express my concern over the proposed uranium mine in our area. We already have one of the highest levels of contamination of radioactive substances in the world. The proposed mine will discharge even more of these pollutants. We currently have no way of properly disposing of the radioactive mine tailings, even more of which will be produced. We have no way to properly safeguard the environment and the health of workers and residents of the area. Furthermore, much of the uranium is destined for export, and we have no guarantee that it will not be used for nuclear weapons. I urge you to hold full public hearings before proceeding further with any development. I await your response.

Here's a letter to a member of the Cabinet:

Dear Sir or Madam:
I am writing to express my concern over the lending record of the World Bank. As you know, in the past, they have supported massive energy, transportation and agricultural projects while ignoring environmental concerns. Although the World Bank has promised new initiatives in environmental policies, it still seems to ignore the idea of energy conservation and alternative sources of energy versus expanded conventional energy production, or rehabilitation of old, smaller irrigation projects versus new mega-projects.

Without consciousness of environmental impact on a worldwide scale, we can expect only further global environmental damage. Our country has an opportunity to influence events. What will our position be?

Although these are sample letters, feel free to copy them or adapt them to your own concerns. Write to any U.S. Senator at: the Senate, Washington, D.C., 20510. Write to any U.S. Congressperson at: the House of Representatives, Washington, D.C. 20515. Write to any Canadian federal politician at: the House of Commons, Parliament Hill, Ottawa, Ontario, K1A 0A6. Canadian residents writing to the House of Commons do not need to stamp the letter.

Take two minutes to write a letter today in support of the environment. If everyone wrote one letter, our politicians and businesses and industries would be inundated with demands for action—and we'd get it.

18

GREEN RESOLUTIONS

It's Easy To Start Feeling Good

We don't need to wait for New Year's Eve to make our environmental resolutions. Spring is an excellent time, with its promise of renewal. The Summer Solstice would also be a good symbolic time to make some commitments to the environment. Fall is like a new year too, with its ingrained "back-to-school" memories, so it's a fine time for resolutions. We're also at the dawn of a new decade and looking toward the millennium—a perfect opportunity to resolve to be kind to our planet. In fact, we can make a promise to ourselves and to the Earth at any time. Here are some ideas.

Spring Resolutions

- Install a water-saving showerhead.
- Give up paper towels for one month.

- Ride your bike to work five times.
- Turn off the tap while you brush your teeth.
- Join an environmental group.
- Plant a tree.
- Give up a piece of lawn to wildflowers or something green that takes less maintenance than lawn.

Summer Resolutions

- Give up chemical pesticides for a month and try some organic gardening.
- Get started on composting or vermicomposting.
- During the good weather set a "no-car" limit of five blocks.
- Put re-refined motor oil in your car, and make sure the old oil is recycled.
- Take your own bags or boxes to the supermarket.
- Eat a vegetarian meal once a week.
- Wash your hands in cold water for a month.

Fall Resolutions

- Use one box of laundry soap instead of detergent.
- Look for last year's leftover school supplies before buying more.
- Photocopy any documents more than one page long on both sides of the paper.
- Take your own coffee cup to work so you won't need to use disposables.
- Wrap a gift in the comics section of the newspaper instead of using new wrapping paper.

- Buy a battery charger and rechargeable batteries to use with the holiday toys.
- Give a green gift to your family and friends.

New Year's Resolutions

- Put a water saver in every toilet in your house or apartment.
- Buy and use at least one screw-in fluorescent lightbulb in place of an incandescent.
- Have your antifreeze recycled.
- Turn down the thermostat five degrees at night.
- Write an environmental protest or suggestion letter to someone in authority.
- Pass on an environmental suggestion to at least two of your friends this year.

Many of these resolutions will take less than two minutes. Some will take a little longer. None will make you break out in a sweat. Nobody is suggesting that you drastically change your lifestyle. Try one or two suggestions to get started. Every positive thing you do will have a positive effect on the planet.

Keep trying the ideas in this book, and remember that we're all in it together. You don't have to spend more than two minutes a day saving the world. But I'm willing to bet that you'll enjoy doing it so much that green thinking will soon become a part of everything you do. Greener Planet acts will become habits, and you'll begin to notice and avoid wasteful and harmful actions.

For every problem, there are opportunities. You can make a real difference with just a small commitment. If you follow any of the suggestions of this book, feel good about yourself. You're doing something positive for the planet and for the rest of us. And for yourself.

Index

acid rain, 50–51, 87
Africa, 219
air conditioning, 52, 53, 62–63, 141, 199; in cars, 64–65; room, 103, 104; car refrigerant, 64–65
Alar, 248
Alberta, 66, 89
alcohol fuel, 65–66, 67–68
all purpose cleaner, 39–40
alternate fuel, 65–66
Amazon, 234, 251
Amchitka, 239
ammonia, 39–40, 170
Antarctic, 52, 181
anti-fouling paints, 204–205
antifreeze, 63–64; dumping of, 63–64
ants, 189, 193–194
aphids, 188, 190–191
appliances, 94–96, 104, 133, 134, 139
Arctic, 52
arsenic, 56–57, 144–145

ashes, 155, 173, 197
Aunt Lenore, 188
Avon, 215–216

bags, 74–75, 113–114; cloth or knitted, 228; net, 231
baking soda, 37–41, 213
Balbina Dam, 88
barbecue, 174, 195; ashes, 191
Barry, 8, 129, 130, 136, 151, 187, 194, 196, 199, 216
base cream, 43–44
batteries, 81, 231
Beluga whales, 88
Berg, Peter, 248
Bhopal, India, 182
bicycle, 56, 58; couriers, 111
biodegradable: cat litter, 213; diapers, 24; paper cups, 112; soaps and cleaners, 46
bioregionalism, 248–249
bird houses, 189, 214, 231
birds, 72, 183, 220; eggs, 183; feeder, 181, 194; hitting win-

water saving showerhead, 131
water supply, 28–30, 34, 64, 87;
    wasted through leaks, 10–11
waterfowl, 209
weatherstripping, 101, 115, 140,
    213
WEB, 250
weeding, 187
wetlands, 87, 88, 240
whales, 220
whitening clothes, 105
wildflowers, 145–146, 198
wildlife, 51, 87, 142–143, 218–
    223

Winnipeg, 150
wood preservatives, 144–145
World Bank, 256
World Energy Congress, 90
World Soc. for the Protection of
    Animals, 72, 221
World Watch Institute, 241–242
World Wildlife Fund Canada,
    142–143, 232, 250–251
worm compost, 125, 177
worms, 157, 167–168, 176–177

Yamaha, 219

zinc, 55